HEINEMANN MODULAR MATHEMATICS

for

EDEXCEL AS AND A-LEVEL

Core Mathematics C4

Greg Attwood Alistair Macpherson Bronwen Moran
Joe Petran Keith Pledger Dave Wilkins

1	Partial fractions	1
2	Coordinate geometry in the (x, y) plane	9
3	The binomial expansion	21
4	Differentiation	33
5	Vectors	47
6	Integration	82

Exam style paper 123
Formulae you need to know 125
List of symbols and notation 126
Answers 129
Index 139

Endorsed by edexcel

heinemann.co.uk
✓ Free online support
✓ Useful weblinks
✓ 24 hour online ordering

01865 888058

Heinemann
Inspiring generations

Heinemann Educational Publishers
Halley Court, Jordan Hill, Oxford OX2 8EJ
Part of Harcourt Education

Heinemann is the registered trademark of
Harcourt Education Limited

First published 2004

09 08 07
10 9 8 7 6

British Library Cataloguing in Publication Data is available
from the British Library on request.

ISBN: 978 0 435511 00 5

Designed by Bridge Creative Services
Typeset by Techset Ltd

Illustrated by Tech-Set Ltd

Cover design by Bridge Creative Services

Printed and bound in China by China Translation & Printing Services Ltd

Acknowledgements
Every effort has been made to contact copyright holders of material reproduced in this book. Any omissions will be rectified in subsequent printings if notice is given to the publishers.

Tel: 01865 888058 email: info.he@heinemann.co.uk

About this book

This book is designed to provide you with the best preparation possible for your Edexcel C4 exam. The authors are members of a senior examining team themselves and have a good understanding of Edexcel's requirements.

Finding your way round

To help you find your way around when you are studying and revising use the:

- **edge colours** – each chapter has a different colour scheme. This helps you to get to the right chapter quickly.
- **contents list** – this lists the headings that identify key syllabus ideas covered in the book so you can turn straight to them. The detailed contents list shows which parts of the C4 syllabus are covered in each section.
- **index** – this lists the headings that identify key syllabus ideas covered in this book so you can turn straight to them.

How sections are structured

- Each section (e.g. 1.1, 1.2) begins with a statement. The statement tells you what is covered in the section.

1.3 You can also split fractions that have more than two linear factors in the denominator into partial fractions.

- Some sections include explanations, which help you understand the maths behind the questions you need to answer in your exam.
- Examples are worked through step-by-step. They are model solutions, as you might write them out. Examiners' hints are given in yellow margin note boxes.
- Each section ends with an exercise, with plenty of questions for practice.

Remembering key ideas

Key ideas you need to remember are listed in a summary of key points at the end of each chapter. The statement at the beginning of a section may be a key point. When key points appear in the teaching, they are marked like this:

■ **The square root of a prime number is a surd.**

Exercises and exam questions

In this book questions are carefully graded so they increase in difficulty and gradually bring you up to standard.

- **Past exam questions** are marked with an *E*.
- **Mixed exercises** at the end of each chapter help you practise answering questions on all the topics you have covered in the chapter.
- **Exam style practice paper** this is designed to help you prepare for the exam itself.
- **Answers** are included at the end of the book – use the answers to check your work.

Contents

About this book v

1 Partial fractions 1
- 1.1 Adding and subtracting algebraic fractions 1
- 1.2 Partial fractions with two linear factors in the denominator 2
- 1.3 Partial fractions with three or more linear factors in the denominator 3
- 1.4 Partial fractions with repeated linear factors in the denominator 5
- 1.5 Improper fractions into partial fractions 6

2 Coordinate geometry in the (x, y) plane 9
- 2.1 Parametric equations used to define the coordinates of a point 9
- 2.2 Using parametric equations in coordinate geometry 11
- 2.3 Converting parametric equations into cartesian equations 14
- 2.4 Finding the area under a curve given by parametric equations 16

3 The binomial expansion 21
- 3.1 The binomial expansion for a positive integral index 21
- 3.2 Using the binomial expansion to expand $(a + bx)^n$ 26
- 3.3 Using partial fractions with the binomial expansion 28

4 Differentiation 33
- 4.1 Differentiating functions given parametrically 33
- 4.2 Differentiating relations which are implicit 35
- 4.3 Differentiating the function a^x 37
- 4.4 Differentiation and rates of change 38
- 4.5 Simple differential equations 39

5 Vectors 47
- 5.1 Vector definitions and vector diagrams 47
- 5.2 Vector arithmetic and the unit vector 51
- 5.3 Using vectors to describe points in 2 or 3 dimensions 55
- 5.4 Cartesian components of a vector in 2 dimensions 56
- 5.5 Cartesian components of a vector in 3 dimensions 59
- 5.6 Extending 2 dimensional vector results to 3 dimensions 62
- 5.7 The scalar product of two vectors 64
- 5.8 The vector equation of a straight line 70
- 5.9 Intersecting straight line vector equations 74
- 5.10 The angle between two straight lines 76

6 Integration 82
- 6.1 Integrating standard functions 82
- 6.2 Integrating using the reverse chain rule 84
- 6.3 Using trigonometric identities in integration 86
- 6.4 Using partial fractions to integrate expressions 89
- 6.5 Using standard patterns to integrate expressions 92
- 6.6 Integration by substitution 95
- 6.7 Integration by parts 99
- 6.8 Numerical integration 102
- 6.9 Integration to find areas and volumes 105
- 6.10 Using integration to solve differential equations 108
- 6.11 Differential equations in context 111

Exam style paper 123

Formulae you need to know 125

List of symbols and notation 126

Answers 129

Index 139

1 Partial fractions

In this chapter you will learn how an algebraic fraction can be written as a sum of two (or more) simpler fractions called partial fractions. You will apply this to areas of calculus and algebra in later chapters.

You learned how to combine simpler fractions into more complex ones in the C3 book, and Section 1.1 reviews this. Splitting complex fractions into partial fractions is the opposite process. Different methods are required depending on the kind of fraction you are starting from.

1.1 You can add (or subtract) two or more simple fractions as long as the denominators are the same.

Example 1

Calculate: **a** $\frac{1}{3}+\frac{3}{8}$ **b** $\frac{2}{x+3}-\frac{1}{(x+1)}$

a $\frac{1}{3}+\frac{3}{8}$

The lowest common multiple of 3 and 8 is 24.

$=\frac{8}{24}+\frac{9}{24}$

Multiply $\frac{1}{3}$ top and bottom by 8 to express in 24ths.
Multiply $\frac{3}{8}$ top and bottom by 3 to express in 24ths.

$=\frac{8+9}{24}$

Add the numerators.

$=\frac{17}{24}$

b $\frac{2}{(x+3)}-\frac{1}{(x+1)}$

The lowest common multiple is $(x+3)(x+1)$, so change both fractions so that the denominators are $(x+3)(x+1)$.

$=\frac{2(x+1)}{(x+3)(x+1)}-\frac{1(x+3)}{(x+3)(x+1)}$

$=\frac{2(x+1)-1(x+3)}{(x+3)(x+1)}$

Subtract the numerators.

$=\frac{2x+2-1x-3}{(x+3)(x+1)}$

Expand the brackets.

$=\frac{x-1}{(x+3)(x+1)}$

Simplify the numerator.

Exercise 1A

Express each of the following as a single fraction:

1 $\frac{1}{3}+\frac{1}{4}$

2 $\frac{3}{4}-\frac{2}{5}$

3 $\frac{3}{x}-\frac{2}{x+1}$

4 $\frac{2}{(x-1)}+\frac{3}{(x+2)}$

5 $\frac{4}{(2x+1)}+\frac{2}{(x-1)}$

6 $\frac{7}{(x-3)}-\frac{2}{(x+4)}$

7 $\frac{3}{2x}-\frac{6}{(x-1)}$

8 $\frac{3}{x}+\frac{2}{(x+1)}+\frac{1}{(x+2)}$

9 $\frac{4}{3x}-\frac{2}{(x-2)}+\frac{1}{(2x+1)}$

10 $\frac{3}{(x-1)}+\frac{2}{(x+1)}+\frac{4}{(x-3)}$

1.2 You can split a fraction with two linear factors in the denominator into partial fractions.

For example, $\frac{x-1}{(x+1)(x+3)}$ is $\frac{2}{(x+3)}-\frac{1}{(x+1)}$ when split into partial fractions (see Example 1).

■ **In general, an expression with two linear terms in the denominator such as $\frac{11}{(x-3)(x+2)}$ can be split into partial fractions of the form $\frac{A}{(x-3)}+\frac{B}{(x+2)}$, where A and B are constants.**

There are two methods of achieving this: by substitution and by equating coefficients.

Example 2

Split $\frac{6x-2}{(x-3)(x+1)}$ into partial fractions by **a** substitution **b** equating coefficients.

a

$$\frac{6x-2}{(x-3)(x+1)} \equiv \frac{A}{(x-3)}+\frac{B}{(x+1)}$$

Set $\frac{6x-2}{(x-3)(x+1)}$ identical to $\frac{A}{(x-3)}+\frac{B}{(x+1)}$.

$$\equiv \frac{A(x+1)+B(x-3)}{(x-3)(x+1)}$$

Add the two fractions.

$$6x-2 \equiv A(x+1)+B(x-3)$$

Because this is an equivalence relation set the numerators equal to each other.

$$6\times 3-2 = A(3+1)+B(3-3)$$
$$16 = 4A$$
$$A = 4$$

To find A substitute $x = 3$.

$$6\times -1-2 = A(-1+1)+B(-1-3)$$
$$-8 = -4B$$
$$B = 2$$

To find B substitute $x = -1$.

$$\therefore \frac{6x-2}{(x-3)(x+1)} \equiv \frac{4}{(x-3)}+\frac{2}{(x+1)}$$

b $\dfrac{6x-2}{(x-3)(x+1)} \equiv \dfrac{A}{(x-3)} + \dfrac{B}{(x+1)}$

Set $\dfrac{6x-2}{(x-3)(x+1)}$ identical to $\dfrac{A}{(x-3)} + \dfrac{B}{(x+1)}$.

$\equiv \dfrac{A(x+1)+B(x-3)}{(x-3)(x+1)}$

Add the two fractions.

$6x - 2 \equiv A(x+1) + B(x-3)$

Because this is an equivalence relation set the numerators equal to each other.

$\equiv Ax + A + Bx - 3B$

Expand the brackets.

$\equiv (A+B)x + (A-3B)$

Collect like terms.

Equate coefficients of x:

$6 = A + B$ ①

Equate constant terms:

$-2 = A - 3B$ ②

You want $(A+B)x + A - 3B = 6x - 2$. Hence coefficient of x is 6, and constant term is -2.

① − ②:

$8 = 4B$

$\Rightarrow \quad B = 2$

Substitute $B = 2$ in ① $\Rightarrow 6 = A + 2$

$A = 4$

Solve simultaneously.

Exercise 1B

1 Express the following as partial fractions:

a $\dfrac{6x-2}{(x-2)(x+3)}$

b $\dfrac{2x+11}{(x+1)(x+4)}$

c $\dfrac{-7x-12}{2x(x-4)}$

d $\dfrac{2x-13}{(2x+1)(x-3)}$

e $\dfrac{6x+6}{x^2-9}$

f $\dfrac{7-3x}{x^2-3x-4}$

g $\dfrac{8-x}{x^2+4x}$

h $\dfrac{2x-14}{x^2+2x-15}$

2 Show that $\dfrac{-2x-5}{(4+x)(2-x)}$ can be written in the form $\dfrac{A}{(4+x)} + \dfrac{B}{(2-x)}$ where A and B are constants to be found.

1.3 You can also split fractions that have more than two linear factors in the denominator into partial fractions.

■ An expression with three or more linear terms in the denominator such as $\dfrac{4}{(x+1)(x-3)(x+4)}$ can be split into $\dfrac{A}{(x+1)} + \dfrac{B}{(x-3)} + \dfrac{C}{(x+4)}$, and so on if there are more terms.

Example 3

Express $\dfrac{6x^2 + 5x - 2}{x(x-1)(2x+1)}$ in partial fractions.

Let $\dfrac{6x^2 + 5x - 2}{x(x-1)(2x+1)} \equiv \dfrac{A}{x} + \dfrac{B}{(x-1)} + \dfrac{C}{(2x+1)}$

The denominators must be x, $(x - 1)$ and $(2x + 1)$.

$\equiv \dfrac{A(x-1)(2x+1) + Bx(2x+1) + Cx(x-1)}{x(x-1)(2x+1)}$

Add the fractions.

$\therefore \quad 6x^2 + 5x - 2$
$\equiv A(x-1)(2x+1) + Bx(2x+1) + Cx(x-1)$

Set the numerators equal.

Let $x = 1$

$6 + 5 - 2 = 0 + B \times 1 \times 3 + 0$

$9 = 3B$

$B = 3$

Let $x = 0$

$0 + 0 - 2 = A \times -1 \times 1 + 0 + 0$

$-2 = -1A$

$A = 2$

Let $x = -\frac{1}{2}$

$\frac{6}{4} - \frac{5}{2} - 2 = 0 + 0 + C \times -\frac{1}{2} \times -\frac{3}{2}$

$-3 = \frac{3}{4}C$

$C = -4$

Proceed by substitution OR by equating coefficients.

So $\dfrac{6x^2 + 5x - 2}{x(x-1)(2x+1)} \equiv \dfrac{2}{x} + \dfrac{3}{x-1} - \dfrac{4}{2x+1}$

Exercise 1C

1 Express the following as partial fractions:

a $\dfrac{2x^2 - 12x - 26}{(x+1)(x-2)(x+5)}$ **b** $\dfrac{-10x^2 - 8x + 2}{x(2x+1)(3x-2)}$ **c** $\dfrac{-5x^2 - 19x - 32}{(x+1)(x+2)(x-5)}$

2 By firstly factorising the denominator, express the following as partial fractions:

a $\dfrac{6x^2 + 7x - 3}{x^3 - x}$ **b** $\dfrac{5x^2 + 15x + 8}{x^3 + 3x^2 + 2x}$ **c** $\dfrac{5x^2 - 15x - 8}{x^3 - 4x^2 + x + 6}$

1.4 You can express a fraction that has repeated linear factors in its denominator as a partial fraction.

■ An expression with repeated linear terms such as $\frac{6x^2 - 29x - 29}{(x+1)(x-3)^2}$ can be split into the form $\frac{A}{(x+1)} + \frac{B}{(x-3)} + \frac{C}{(x-3)^2}$.

Example 4

Split $\frac{11x^2 + 14x + 5}{(x+1)^2(2x+1)}$ into partial fraction form.

Let $\frac{11x^2 + 14x + 5}{(x+1)^2(2x+1)} \equiv \frac{A}{(x+1)} + \frac{B}{(x+1)^2} + \frac{C}{(2x+1)}$ — You need denominators of $(x + 1)$, $(x + 1)^2$ and $(2x + 1)$.

$\equiv \frac{A(x+1)(2x+1) + B(2x+1) + C(x+1)^2}{(x+1)^2(2x+1)}$ — Add the three fractions.

Hence $11x^2 + 14x + 5$
$\equiv A(x+1)(2x+1) + B(2x+1) + C(x+1)^2$ — Set the numerators equal.

Let $x = -1$

$11 - 14 + 5 = A \times 0 + B \times -1 + C \times 0$ — To find B substitute $x = -1$.

$2 = -1B$

$B = -2$

Let $x = -\frac{1}{2}$

$\frac{11}{4} - 7 + 5 = A \times 0 + B \times 0 + C \times \frac{1}{4}$ — To find C substitute $x = -\frac{1}{2}$.

$\frac{3}{4} = \frac{1}{4}C$

$C = 3$

$11 = 2A + C$ — Equate terms in x^2. Term in x^2 is $A \times 2x^2 + C \times x^2$.

$11 = 2A + 3$ — Substitute $C = 3$.

$2A = 8$

$A = 4$

Hence $\frac{11x^2 + 14x + 5}{(x+1)^2(2x+1)}$

$\equiv \frac{4}{(x+1)} - \frac{2}{(x+1)^2} + \frac{3}{(2x+1)}$

Exercise 1D

Put the following into partial fraction form:

1 $\frac{3x^2 + x + 2}{x^2(x+1)}$

2 $\frac{-x^2 - 10x - 5}{(x+1)^2(x-1)}$

3 $\frac{2x^2 + 2x - 18}{x(x-3)^2}$

4 $\frac{7x^2 - 42x + 64}{x(x-4)^2}$

5 $\frac{5x^2 - 2x - 1}{x^3 - x^2}$

6 $\frac{2x^2 + 2x - 18}{x^3 - 6x^2 + 9x}$

7 $\frac{2x}{(x+2)^2}$

8 $\frac{x^2 + 5x + 7}{(x+2)^3}$

1.5 You can split improper fractions into partial fractions by dividing the numerator by the denominator.

■ **An algebraic fraction is improper when the degree of the numerator is equal to, or larger than, the degree of the denominator. An improper fraction must be divided first to obtain a number and a proper fraction before it can be expressed in partial fractions.**

$\frac{x^2}{x(x-3)}$, $\frac{x^3+4x^2+2}{(x+1)(x-3)}$ and $\frac{x^4}{(x-1)^2(x+2)}$ are all examples of improper fractions.

Example 5

Express $\frac{3x^2-3x-2}{(x-1)(x-2)}$ in partial fractions.

$$\frac{3x^2-3x-2}{(x-1)(x-2)} \equiv \frac{3x^2-3x-2}{x^2-3x+2}$$

Multiply out the denominator.

Divide the denominator into the numerator.

$$\equiv x^2-3x+2\overline{\smash{)}3x^2-3x-2}\quad\text{quotient } 3$$
$$\begin{array}{r} 3x^2-9x+6 \\ \hline 6x-8 \end{array}$$

It goes in 3 times, with a remainder of $6x-8$.

Therefore

$$\frac{3x^2-3x-2}{(x-1)(x-2)} \equiv 3 + \frac{6x-8}{x^2-3x+2}$$

Write $\frac{3x^2-3x-2}{(x-1)(x-2)}$ as a mixed number fraction.

$$= 3 + \frac{6x-8}{(x-1)(x-2)}$$

Factorise x^2-3x+2.

Let $\frac{6x-8}{(x-1)(x-2)} \equiv \frac{A}{(x-1)} + \frac{B}{(x-2)}$

The denominators must be $(x-1)$ and $(x-2)$.

$$\equiv \frac{A(x-2)+B(x-1)}{(x-1)(x-2)}$$

Add the two fractions.

$$6x-8 \equiv A(x-2)+B(x-1)$$

Set the numerators equal.

Let $x=2$, $12-8 = A \times 0 + B \times 1$

$B = 4$

Substitute $x = 2$ to find B.

Let $x=1$, $6-8 = A \times -1 + B \times 0$

$A = 2$

Substitute $x = 1$ to find A.

$$\frac{3x^2-3x-2}{(x-1)(x-2)} = 3 + \frac{6x-8}{(x-1)(x-2)}$$

$$\equiv 3 + \frac{2}{(x-1)} + \frac{4}{(x-2)}$$

Write out full solution.

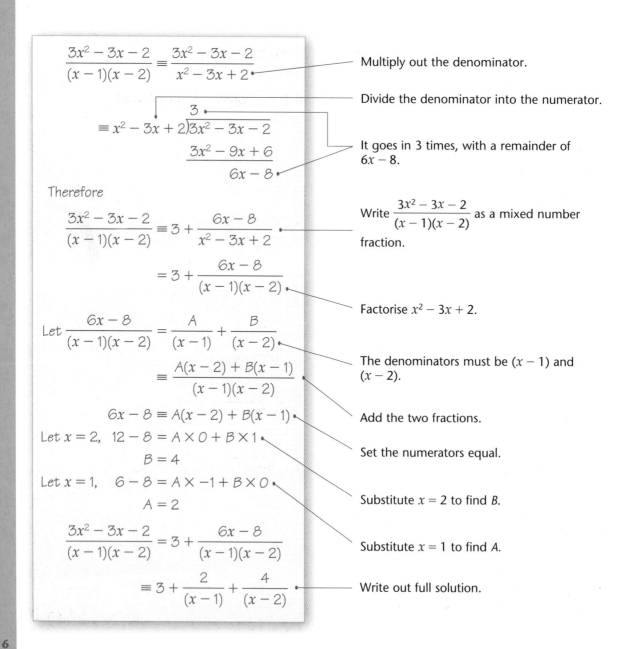

Exercise 1E

1 Express the following improper fractions as a partial fraction:

a $\dfrac{x^2+3x-2}{(x+1)(x-3)}$

b $\dfrac{x^2-10}{(x-2)(x+1)}$

c $\dfrac{x^3-x^2-x-3}{x(x-1)}$

d $\dfrac{2x^2-1}{(x+1)^2}$

2 By factorising the denominator, express the following as partial fraction:

a $\dfrac{4x^2+17x-11}{x^2+3x-4}$

b $\dfrac{x^4-4x^3+9x^2-17x+12}{x^3-4x^2+4x}$

3 Show that $\dfrac{-3x^3-4x^2+19x+8}{x^2+2x-3}$ can be expressed in the form

$A+Bx+\dfrac{C}{(x-1)}+\dfrac{D}{(x+3)}$, where A, B, C and D are constants to be found.

Mixed exercise 1F

1 Express the following as a partial fraction:

a $\dfrac{x-3}{x(x-1)}$

b $\dfrac{7x^2+2x-2}{x^2(x+1)}$

c $\dfrac{-15x+21}{(x-2)(x+1)(x-5)}$

d $\dfrac{x^2+1}{x(x-2)}$

2 Write the following algebraic fractions as a partial fraction:

a $\dfrac{3x+1}{x^2+2x+1}$

b $\dfrac{2x^2+2x-8}{x^2+2x-3}$

c $\dfrac{3x^2+12x+8}{(x+2)^3}$

d $\dfrac{x^4}{x^2-2x+1}$

3 Given that $f(x)=2x^3+9x^2+10x+3$:

a Show that -3 is a root of $f(x)$.

b Express $\dfrac{10}{f(x)}$ as partial fractions.

E

(adapted)

Summary of key points

1 An algebraic fraction can be written as a sum of two or more simpler fractions. This technique is called splitting into partial fractions.

2 An expression with two linear terms in the denominator such as $\frac{11}{(x-3)(x+2)}$ can be split by converting into the form $\frac{A}{(x-3)} + \frac{B}{(x+2)}$.

3 An expression with three or more linear terms such as $\frac{4}{(x+1)(x-3)(x+4)}$ can be split by converting into the form $\frac{A}{(x+1)} + \frac{B}{(x-3)} + \frac{C}{(x+4)}$ and so on if there are more terms.

4 An expression with repeated terms in the denominator such as $\frac{6x^2 - 29x - 29}{(x+1)(x-3)^2}$ can be split by converting into the form $\frac{A}{(x+1)} + \frac{B}{(x-3)} + \frac{C}{(x-3)^2}$.

5 An improper fraction is one where the index of the numerator is equal to or higher than the index of the denominator. An improper fraction must be divided first to obtain a number and a proper fraction before you can express it in partial fractions.

- For example, $\frac{x^2+3x+4}{x^2+3x+2} = 1 + \frac{2}{x^2+3x+2} = 1 + \frac{A}{(x+1)} + \frac{B}{(x+2)}$.

2 Coordinate geometry in the (x, y) plane

In this chapter you will learn how to solve problems involving parametric equations.

2.1 You can define the coordinates of a point on a curve using parametric equations. In parametric equations, coordinates x and y are expressed as $x = f(t)$ and $y = g(t)$, where the variable t is a parameter.

Example 1

Draw the curve given by the parametric equations $x = 2t$, $y = t^2$, for $-3 \leqslant t \leqslant 3$.

t	-3	-2	-1	0	1	2	3
$x = 2t$	-6	-4	-2	0	2	4	6
$y = t^2$	9	4	1	0	1	4	9

Draw a table to show values of t, x and y. Choose values for t. Here $-3 \leqslant t \leqslant 3$.

Work out the value of x and the value of y for each value of t by substituting values of t into the parametric equations $x = 2t$ and $y = t^2$.

e.g. for $t = 2$:

$x = 2t = 2(2) = 4$ $\quad$ $y = t^2 = (2)^2 = 4$

So when $t = 2$ the curve passes through the point (4, 4).

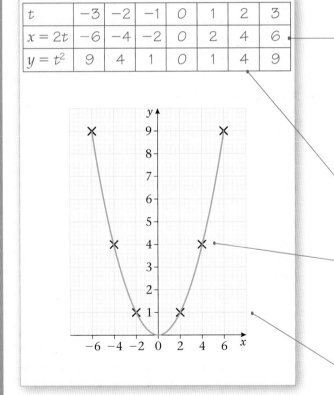

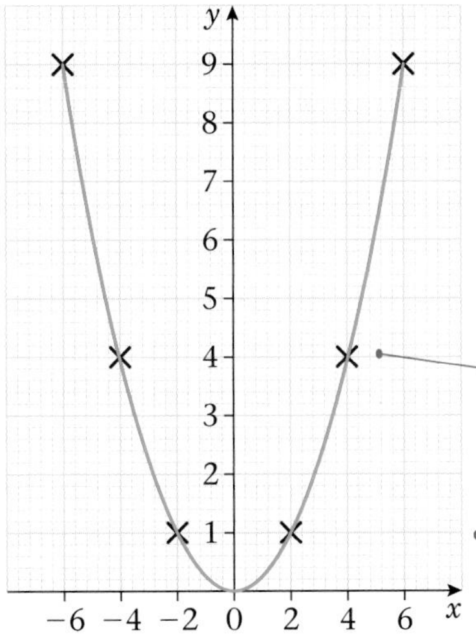

Plot the points $(-6, 9)$, $(-4, 4)$, $(-2, 1)$, $(0, 0)$, $(2, 1)$, $(4, 4)$, $(6, 9)$ and draw the graph through the points.

Example 2

A curve has parametric equations $x = 2t$, $y = t^2$. Find the cartesian equation of the curve.

$x = 2t$

So $t = \dfrac{x}{2}$ ①

$y = t^2$ ②

Substitute ① into ②:

$y = \left(\dfrac{x}{2}\right)^2$

The cartesian equation is $y = \dfrac{x^2}{4}$.

A cartesian equation is an equation in terms of x and y only.

To obtain the cartesian equation, eliminate t from the parametric equations $x = 2t$ and $y = t^2$.

Rearrange $x = 2t$ for t.
Divide each side by 2.

Substitute $t = \dfrac{x}{2}$ into $y = t^2$.

Expand the brackets, so that $\left(\dfrac{x}{2}\right)^2 = \dfrac{x}{2} \times \dfrac{x}{2} = \dfrac{x^2}{4}$

Exercise 2A

1 A curve is given by the parametric equations $x = 2t$, $y = \dfrac{5}{t}$ where $t \neq 0$. Complete the table and draw a graph of the curve for $-5 \leqslant t \leqslant 5$.

t	-5	-4	-3	-2	-1	-0.5	0.5	1	2	3	4	5
$x = 2t$	-10	-8				-1						
$y = \dfrac{5}{t}$	-1	-1.25					10					

2 A curve is given by the parametric equations $x = t^2$, $y = \dfrac{t^3}{5}$. Complete the table and draw a graph of the curve for $-4 \leqslant t \leqslant 4$.

t	-4	-3	-2	-1	0	1	2	3	4
$x = t^2$	16								
$y = \dfrac{t^3}{5}$	-12.8								

3 Sketch the curves given by these parametric equations:

a $x = t - 2$, $y = t^2 + 1$ for $-4 \leqslant t \leqslant 4$
b $x = t^2 - 2$, $y = 3 - t$ for $-3 \leqslant t \leqslant 3$
c $x = t^2$, $y = t(5 - t)$ for $0 \leqslant t \leqslant 5$
d $x = 3\sqrt{t}$, $y = t^3 - 2t$ for $0 \leqslant t \leqslant 2$
e $x = t^2$, $y = (2 - t)(t + 3)$ for $-5 \leqslant t \leqslant 5$

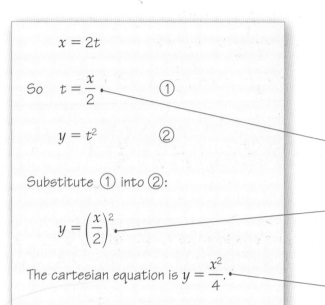

4 Find the cartesian equation of the curves given by these parametric equations:

a $x = t - 2,\ y = t^2$

b $x = 5 - t,\ y = t^2 - 1$

c $x = \frac{1}{t},\ y = 3 - t,\ t \neq 0$

d $x = 2t + 1,\ y = \frac{1}{t},\ t \neq 0$

e $x = 2t^2 - 3,\ y = 9 - t^2$

f $x = \sqrt{t},\ y = t(9 - t)$

g $x = 3t - 1,\ y = (t - 1)(t + 2)$

h $x = \frac{1}{t - 2},\ y = t^2,\ t \neq 2$

i $x = \frac{1}{t + 1},\ y = \frac{1}{t - 2},\ t \neq -1,\ t \neq 2$

j $x = \frac{t}{2t - 1},\ y = \frac{t}{t + 1},\ t \neq -1,\ t \neq \frac{1}{2}$

5 Show that the parametric equations:

i $x = 1 + 2t,\ y = 2 + 3t$

ii $x = \frac{1}{2t - 3},\ y = \frac{t}{2t - 3},\ t \neq \frac{3}{2}$

represent the same straight line.

2.2 You need to be able to use parametric equations to solve problems in coordinate geometry.

Example 3

The diagram shows a sketch of the curve with parametric equations $x = t - 1,\ y = 4 - t^2$. The curve meets the x-axis at the points A and B. Find the coordinates of A and B.

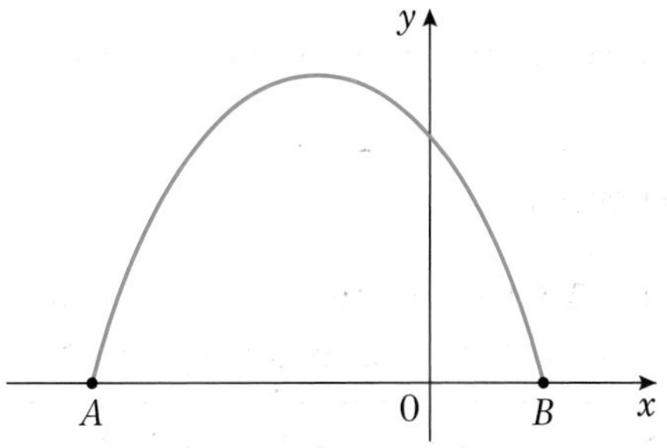

①	$y = 4 - t^2$
Substitute	$y = 0$
	$4 - t^2 = 0$
	$t^2 = 4$
So	$t = \pm 2$
②	$x = t - 1$
Substitute	$t = 2$
	$x = (2) - 1$
	$= 1$
Substitute	$t = -2$
	$x = (-2) - 1$
	$= -3$
The coordinates of A and B are $(-3, 0)$	
and $(1, 0)$.	

Find the values of t at A and B.

The curve meets the x-axis when $y = 0$, so substitute $y = 0$ into $y = 4 - t^2$ and solve for t.

Take the square root of each side. Remember there are two solutions when you take a square root.

Find the value of x at A and B. Substitute $t = 2$ and $t = -2$ into $x = t - 1$.

Example 4

A curve has parametric equations $x = at$, $y = a(t^3 + 8)$, where a is a constant. The curve passes through the point $(2, 0)$. Find the value of a.

① $\qquad y = a(t^3 + 8)$
Substitute $\quad y = 0$
$\qquad a(t^3 + 8) = 0$
$\qquad t^3 + 8 = 0$
$\qquad t^3 = -8$
So $\qquad t = -2$
② $\qquad x = at$
Substitute $\quad x = 2$ and $t = -2$
$\qquad a(-2) = 2$
So $\qquad a = -1$
(The parametric equations of the curve
are $x = -t$ and $y = -(t^3 + 8)$.)

The curve passes through $(2, 0)$, so there is a value of t for which $x = 2$ and $y = 0$.

Find t. Substitute $y = 0$ into $y = a(t^3 + 8)$ and solve for t.

Divide each side by a.

Take the cube root of each side. $\sqrt[3]{-8} = -2$.

Find a. Substitute $x = 2$ and $t = -2$ into $x = at$.

Divide each side by -2.

Example 5

A curve is given parametrically by the equations $x = t^2$, $y = 4t$. The line $x + y + 4 = 0$ meets the curve at A. Find the coordinates of A.

① $\quad x + y + 4 = 0$
Substitute
$\qquad t^2 + 4t + 4 = 0$
$\qquad (t + 2)^2 = 0$
$\qquad t + 2 = 0$
So $\qquad t = -2$
Substitute $\quad t = -2$
② $\qquad x = t^2$
$\qquad = (-2)^2$
$\qquad = 4$
③ $\qquad y = 4t$
$\qquad = 4(-2)$
$\qquad = -8$
The coordinates of A are $(4, -8)$.

Find the value of t at A.
Solve the equations simultaneously.
Substitute $x = t^2$ and $y = 4t$ into $x + y + 4 = 0$.

Factorise.

Take the square root of each side.

Find the coordinates of A.
Substitute $t = -2$ into the parametric equations.

Exercise 2B

1 Find the coordinates of the point(s) where the following curves meet the x-axis:

a $x = 5 + t$, $y = 6 - t$

b $x = 2t + 1$, $y = 2t - 6$

c $x = t^2$, $y = (1 - t)(t + 3)$

d $x = \frac{1}{t}$, $y = \sqrt{(t - 1)(2t - 1)}$, $t \neq 0$

e $x = \frac{2t}{1 + t}$, $y = t - 9$, $t \neq -1$

2 Find the coordinates of the point(s) where the following curves meet the y-axis:

a $x = 2t$, $y = t^2 - 5$

b $x = \sqrt{(3t - 4)}$, $y = \frac{1}{t^2}$, $t \neq 0$

c $x = t^2 + 2t - 3$, $y = t(t - 1)$

d $x = 27 - t^3$, $y = \frac{1}{t - 1}$, $t \neq 1$

e $x = \frac{t - 1}{t + 1}$, $y = \frac{2t}{t^2 + 1}$, $t \neq -1$

3 A curve has parametric equations $x = 4at^2$, $y = a(2t - 1)$, where a is a constant. The curve passes through the point $(4, 0)$. Find the value of a.

4 A curve has parametric equations $x = b(2t - 3)$, $y = b(1 - t^2)$, where b is a constant. The curve passes through the point $(0, -5)$. Find the value of b.

5 A curve has parametric equations $x = p(2t - 1)$, $y = p(t^3 + 8)$, where p is a constant. The curve meets the x-axis at $(2, 0)$ and the y-axis at A.

a Find the value of p.

b Find the coordinates of A.

6 A curve is given parametrically by the equations $x = 3qt^2$, $y = 4(t^3 + 1)$, where q is a constant. The curve meets the x-axis at X and the y-axis at Y. Given that $OX = 2OY$, where O is the origin, find the value of q.

7 Find the coordinates of the point of intersection of the line with parametric equations $x = 3t + 2$, $y = 1 - t$ and the line $y + x = 2$.

8 Find the coordinates of the points of intersection of the curve with parametric equations $x = 2t^2 - 1$, $y = 3(t + 1)$ and the line $3x - 4y = 3$.

9 Find the values of t at the points of intersection of the line $4x - 2y - 15 = 0$ with the parabola $x = t^2$, $y = 2t$ and give the coordinates of these points.

10 Find the points of intersection of the parabola $x = t^2$, $y = 2t$ with the circle $x^2 + y^2 - 9x + 4 = 0$.

2.3 You need to be able to convert parametric equations into a cartesian equation.

Example 6

A curve has parametric equations $x = \sin t + 2$, $y = \cos t - 3$.

a Find the cartesian equation of the curve. **b** Draw a graph of the curve.

a $x = \sin t + 2$

So $\sin t = x - 2$

$y = \cos t - 3$

So $\cos t = y + 3$

As $\sin^2 t + \cos^2 t = 1$,

the cartesian equation of the curve is $(x - 2)^2 + (y + 3)^2 = 1$

Eliminate t from the parametric equations $x = \sin t + 2$ and $y = \cos t - 3$.

Remember $\sin^2 t + \cos^2 t = 1$.

Rearrange $x = \sin t + 2$ for $\sin t$. Take 2 from each side.

Rearrange $y = \cos t - 3$ for $\cos t$. Add 3 to each side.

Square $\sin t$ and $\cos t$ so that
$\sin^2 t = (x - 2)^2$
$\cos^2 t = (y + 3)^2$.

Remember $(x - a)^2 + (y - b)^2 = r^2$ is the equation of a circle centre (a, b), radius r.

b

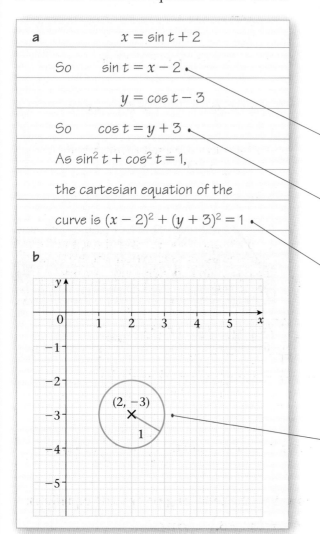

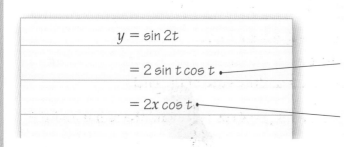

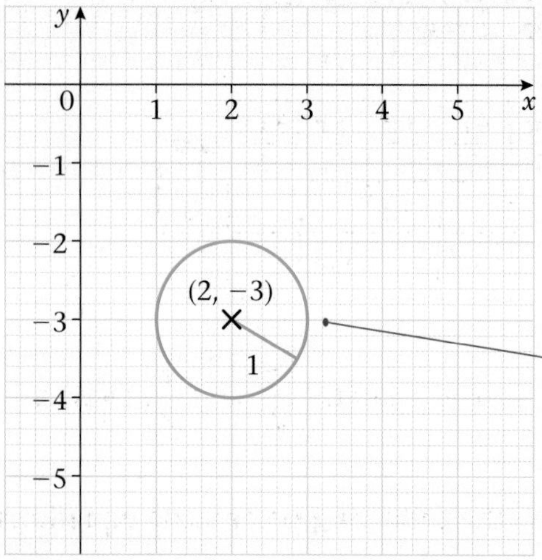

Compare $(x - 2)^2 + (y + 3)^2 = 1$ with $(x - a)^2 + (y - b)^2 = r^2$.
Here $a = 2$, $b = -3$ and $r = 1$.

So $(x - 2)^2 + (y + 3)^2 = 1$ is a circle centre $(2, -3)$ and radius 1.

Example 7

A curve has parametric equations $x = \sin t$, $y = \sin 2t$. Find the cartesian equation of the curve.

Eliminate t between the parametric equations $x = \sin t$ and $y = \sin 2t$.

$y = \sin 2t$

$= 2 \sin t \cos t$

$= 2x \cos t$

Remember $\sin 2t = 2 \sin t \cos t$.

$x = \sin t$, so replace $\sin t$ by x in $y = 2 \sin t \cos t$.

$\sin^2 t + \cos^2 t = 1$

Find $\cos t$ in terms of x.
Rearrange $\sin^2 t + \cos^2 t = 1$ for $\cos t$.

$\cos^2 t = 1 - \sin^2 t$

Take $\sin^2 t$ from each side.

$\cos^2 t = 1 - x^2$

$x = \sin t$, so replace $\sin t$ by x in $\cos^2 t = 1 - \sin^2 t$.

$\cos t = \sqrt{1 - x^2}$

Take the square roof of each side.

So $y = 2x\sqrt{1 - x^2}$

Replace $\cos t$ by $\sqrt{1 - x^2}$ in $y = 2x \cos t$.

This equation can also be written in the form $y^2 = 4x^2(1 - x^2)$.

Exercise 2C

You are unlikely to be asked this kind of question in your exam. However, here it will help your understanding of parametric equations.

1 A curve is given by the parametric equations $x = 2 \sin t$, $y = \cos t$.
Complete the table and draw a graph of the curve for $0 \leqslant t \leqslant 2\pi$.

t	0	$\frac{\pi}{6}$	$\frac{\pi}{3}$	$\frac{\pi}{2}$	$\frac{2\pi}{3}$	$\frac{5\pi}{6}$	π	$\frac{7\pi}{6}$	$\frac{4\pi}{3}$	$\frac{3\pi}{2}$	$\frac{5\pi}{3}$	$\frac{11\pi}{6}$	2π
$x = 2 \sin t$			1.73		1.73			−1		−2			0
$y = \cos t$		0.87					−1		−0.5		0.5		

2 A curve is given by the parametric equations $x = \sin t$, $y = \tan t$, $-\frac{\pi}{2} < t < \frac{\pi}{2}$. Draw a graph of the curve.

3 Find the cartesian equation of the curves given by the following parametric equations:

a $x = \sin t$, $y = \cos t$

b $x = \sin t - 3$, $y = \cos t$

c $x = \cos t - 2$, $y = \sin t + 3$

d $x = 2 \cos t$, $y = 3 \sin t$

e $x = 2 \sin t - 1$, $y = 5 \cos t + 4$

f $x = \cos t$, $y = \sin 2t$

g $x = \cos t$, $y = 2 \cos 2t$

h $x = \sin t$, $y = \tan t$

i $x = \cos t + 2$, $y = 4 \sec t$

j $x = 3 \cot t$, $y = \operatorname{cosec} t$

4 A circle has parametric equations $x = \sin t - 5$, $y = \cos t + 2$.

a Find the cartesian equation of the circle.

b Write down the radius and the coordinates of the centre of the circle.

5 A circle has parametric equations $x = 4 \sin t + 3$, $y = 4 \cos t - 1$. Find the radius and the coordinates of the centre of the circle.

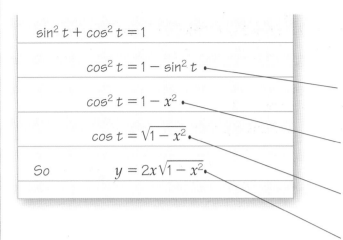

2.4 You need to be able to find the area under a curve given by parametric equations.

■ **The area under a graph is given by $\int y\,dx$. By the chain rule $\int y\,dx = \int y\frac{dx}{dt}\,dt$.**

Example 8

A curve has parametric equations $x = 5t^2$, $y = t^3$. Work out $\int_1^2 y\frac{dx}{dt}\,dt$.

$y\frac{dx}{dt} = t^3\frac{dx}{dt}$

Find $y\frac{dx}{dt}$. Substitute $y = t^3$.

$\frac{dx}{dt} = \frac{d}{dt}(5t^2)$

$= 10t$

Work out $\frac{dx}{dt}$. Here $x = 5t^2$.

$\frac{d}{dt}(5t^2) = 5 \times 2t^{2-1}$

$= 10t$

So $y\frac{dx}{dt} = t^3 \times 10t$

$= 10t^4$

Simplify the expression so that

$t^3 \times 10t = 10t^{3+1}$

$= 10t^4$

$\int_1^2 y\frac{dx}{dt}\,dt = \int_1^2 10t^4\,dt$

$= \left[2t^5\right]_1^2$

Integrate, so that

$\int 10t^4\,dt = \frac{10}{4+1}t^{4+1}$

$= \frac{10}{5}t^5$

$= 2t^5$

$= 2(2)^5 - 2(1)^5$

$= 64 - 2$

$= 62$

Work out $\left[2t^5\right]_1^2$. Substitute $t = 2$ and $t = 1$ into $2t^5$ and subtract.

Example 9

The diagram shows a sketch of the curve with parametric equations $x = t^2$, $y = 2t(3 - t)$, $t \geqslant 0$. The curve meets the x-axis at $x = 0$ and $x = 9$. The shaded region R is bounded by the curve and the x-axis.

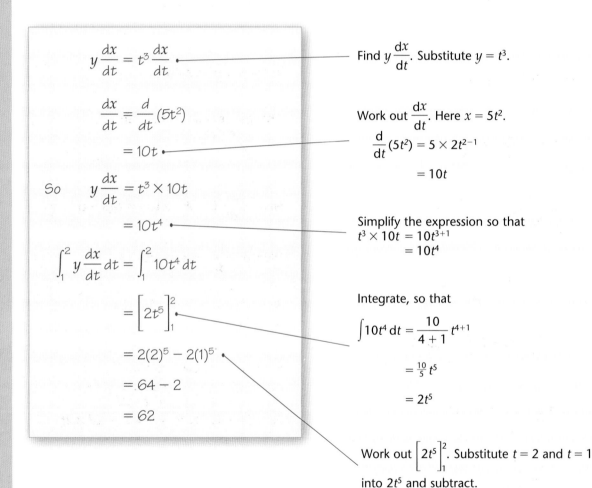

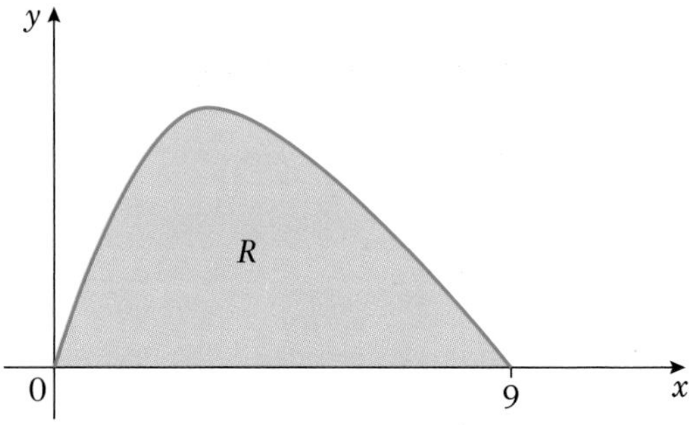

a Find the value of t when

i $x = 0$ **ii** $x = 9$

b Find the area of R.

a i $x = t^2$

$t^2 = 0$ — Substitute $x = 0$ into $x = t^2$. Take the square root of each side.

So $t = 0$

ii $x = t^2$

$t^2 = 9$ — Substitute $x = 9$ into $x = t^2$. Take the square root of each side. $\sqrt{9} = \pm 3$.

So $t = 3$ — Ignore $t = -3$ as $t \geqslant 0$.

b Area of $R = \int_0^9 y\,dx$ — Integrate parametrically.

$= \int_0^3 y \frac{dx}{dt}\,dt$ — Change the limits of the integral. $t = 0$ when $x = 0$; $t = 3$ when $x = 9$

$= \int_0^3 2t(3 - t)\frac{dx}{dt}\,dt$ — Find $\int y \frac{dx}{dt}\,dt$. Substitute $y = 2t(3 - t)$.

$= \int_0^3 2t(3 - t) \times 2t\,dt$ — Work out $\frac{dx}{dt}$. Here $x = t^2$.

$$\frac{dx}{dt} = \frac{d}{dt}(t^2) = 2t$$

$= \int_0^3 (6t - 2t^2) \times 2t\,dt$

$= \int_0^3 12t^2 - 4t^3\,dt$

Expand the brackets, so that

① $2t(3 - t) = 2t \times 3 - 2t \times t = 6t - 2t^2$

② $(6t - 2t^2) \times 2t = 6t \times 2t - 2t^2 \times 2t = 12t^2 - 4t^3$

$= \left[4t^3 - t^4\right]_0^3$

Integrate each term, so that

① $\int 12t^2\,dt = \frac{12}{3}t^{2+1} = 4t^3$

② $\int 4t^3\,dt = \frac{4}{4}t^{3+1} = t^4$

$= [4(3)^3 - (3)4] - [4(0)^3 - (0)^4]$

$= (108 - 81) - (0 - 0)$

$= 27$

The area of $R = 27$.

Work out $\left[4t^3 - t^4\right]_0^3$. Substitute $t = 3$ and $t = 0$ into $4t^3 - t^4$ and subtract.

Exercise 2D

1 The following curves are given parametrically. In each case, find an expression for $y\dfrac{dx}{dt}$ in terms of t.

a $x = t + 3,\ y = 4t - 3$

b $x = t^3 + 3t,\ y = t^2$

c $x = (2t - 3)^2,\ y = 1 - t^2$

d $x = 6 - \dfrac{1}{t},\ y = 4t^3,\ t > 0$

e $x = \sqrt{t},\ y = 6t^3,\ t \geq 0$

f $x = \dfrac{4}{t^2},\ y = 5t^2,\ t < 0$

g $x = 5t^{\frac{1}{2}},\ y = 4t^{-\frac{3}{2}},\ t > 0$

h $x = t^{\frac{1}{3}} - 1,\ y = \sqrt{t},\ t \geq 0$

i $x = 16 - t^4,\ y = 3 - \dfrac{2}{t},\ t < 0$

j $x = 6t^{\frac{2}{3}},\ y = t^2$

2 A curve has parametric equations $x = 2t - 5,\ y = 3t + 8$. Work out $\displaystyle\int_0^4 y\frac{dx}{dt}\,dt$.

3 A curve has parametric equations $x = t^2 - 3t + 1,\ y = 4t^2$. Work out $\displaystyle\int_{-1}^5 y\frac{dx}{dt}\,dt$.

4 A curve has parametric equations $x = 3t^2,\ y = \dfrac{1}{t} + t^3,\ t > 0$. Work out $\displaystyle\int_{0.5}^3 y\frac{dx}{dt}\,dt$.

5 A curve has parametric equations $x = t^3 - 4t,\ y = t^2 - 1$. Work out $\displaystyle\int_{-2}^2 y\frac{dx}{dt}\,dt$.

6 A curve has parametric equations $x = 9t^{\frac{4}{3}},\ y = t^{-\frac{1}{3}},\ t > 0$.

a Show that $y\dfrac{dx}{dt} = a$, where a is a constant to be found.

b Work out $\displaystyle\int_3^5 y\frac{dx}{dt}\,dt$.

7 A curve has parametric equations $x = \sqrt{t},\ y = 4\sqrt{t^3},\ t > 0$.

a Show that $y\dfrac{dx}{dt} = pt$, where p is a constant to be found.

b Work out $\displaystyle\int_1^6 y\frac{dx}{dt}\,dt$.

8 The diagram shows a sketch of the curve with parametric equations $x = t^2 - 3,\ y = 3t$, $t > 0$. The shaded region R is bounded by the curve, the x-axis and the lines $x = 1$ and $x = 6$.

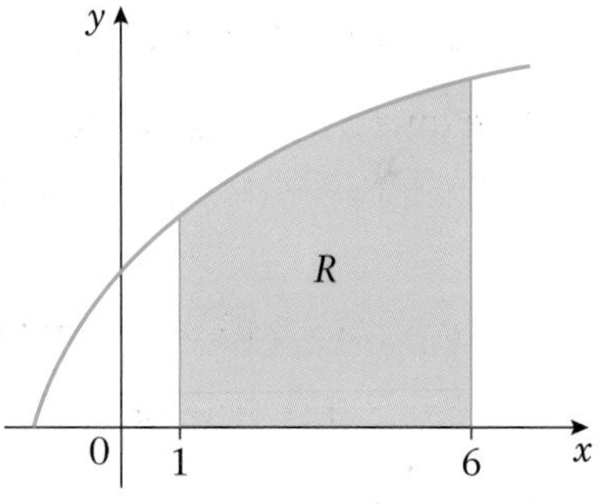

a Find the value of t when

i $x = 1$

ii $x = 6$

b Find the area of R.

9 The diagram shows a sketch of the curve with parametric equations $x = 4t^2$, $y = t(5 - 2t)$, $t \geqslant 0$. The shaded region R is bounded by the curve and the x-axis. Find the area of R.

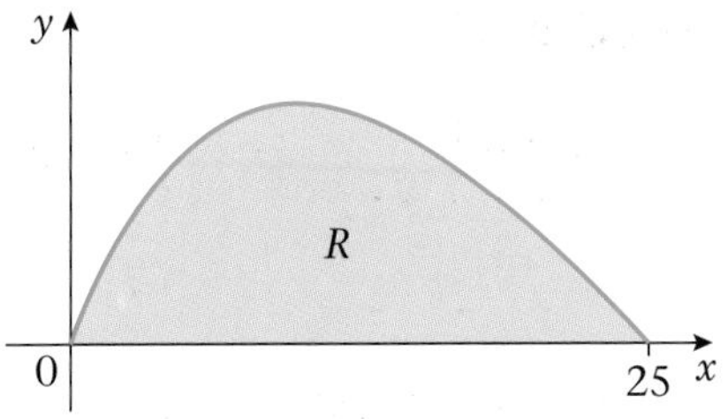

10 The region R is bounded by the curve with parametric equations $x = t^3$, $y = \frac{1}{3t^2}$, the x-axis and the lines $x = -1$ and $x = -8$.

a Find the value of t when

i $x = -1$ **ii** $x = -8$

b Find the area of R.

Mixed exercise 2E

1 The diagram shows a sketch of the curve with parametric equations $x = 4\cos t$, $y = 3\sin t$, $0 \leqslant t < 2\pi$.

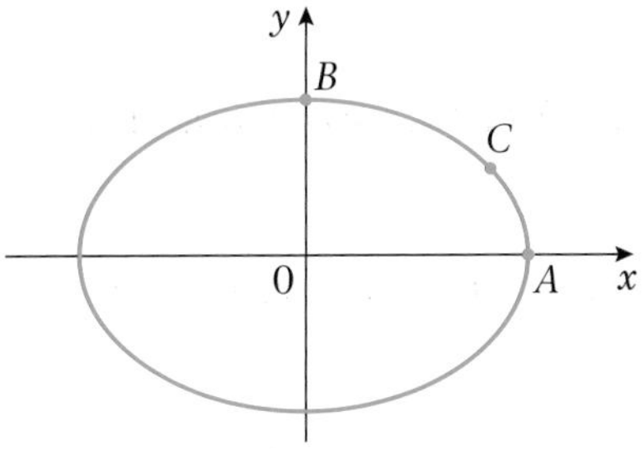

a Find the coordinates of the points A and B.

b The point C has parameter $t = \frac{\pi}{6}$. Find the exact coordinates of C.

c Find the cartesian equation of the curve.

2 The diagram shows a sketch of the curve with parametric equations $x = \cos t$, $y = \frac{1}{2}\sin 2t$.

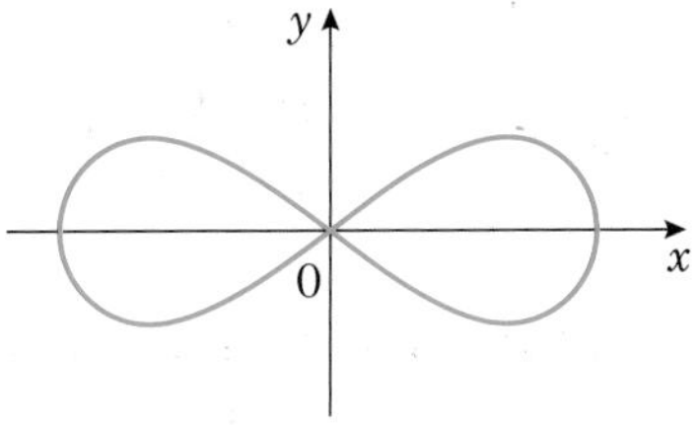

$0 \leqslant t < 2\pi$. The curve is symmetrical about both axes.

a Copy the diagram and label the points having parameters $t = 0$, $t = \frac{\pi}{2}$, $t = \pi$ and $t = \frac{3\pi}{2}$.

b Show that the cartesian equation of the curve is $y^2 = x^2(1 - x^2)$.

3 A curve has parametric equations $x = \sin t$, $y = \cos 2t$, $0 \leqslant t < 2\pi$.

a Find the cartesian equation of the curve.

The curve cuts the x-axis at $(a, 0)$ and $(b, 0)$.

b Find the value of a and b.

4 A curve has parametric equations $x = \frac{1}{1 + t}$, $y = \frac{1}{(1 + t)(1 - t)}$, $t \neq \pm 1$.

Express t in terms of x. Hence show that the cartesian equation of the curve is $y = \frac{x^2}{2x - 1}$.

5 A circle has parametric equations $x = 4\sin t - 3$, $y = 4\cos t + 5$.

a Find the cartesian equation of the circle.

b Draw a sketch of the circle.

c Find the exact coordinates of the points of intersection of the circle with the y-axis.

6 Find the cartesian equation of the line with parametric equations $x = \dfrac{2 - 3t}{1 + t}$, $y = \dfrac{3 + 2t}{1 + t}$, $t \neq -1$.

7 A curve has parametric equations $x = t^2 - 1$, $y = t - t^3$, where t is a parameter.

a Draw a graph of the curve for $-2 \leqslant t \leqslant 2$.

b Find the area of the finite region enclosed by the loop of the curve.

8 A curve has parametric equations $x = t^2 - 2$, $y = 2t$, where $-2 \leqslant t \leqslant 2$.

a Draw a graph of the curve.

b Indicate on your graph where **i** $t = 0$ **ii** $t > 0$ **iii** $t < 0$

c Calculate the area of the finite region enclosed by the curve and the y-axis.

9 Find the area of the finite region bounded by the curve with parametric equations $x = t^3$, $y = \dfrac{4}{t}$, $t \neq 0$, the x-axis and the lines $x = 1$ and $x = 8$.

10 The diagram shows a sketch of the curve with parametric equations $x = 3\sqrt{t}$, $y = t(4 - t)$, where $0 \leqslant t \leqslant 4$. The region R is bounded by the curve and the x-axis.

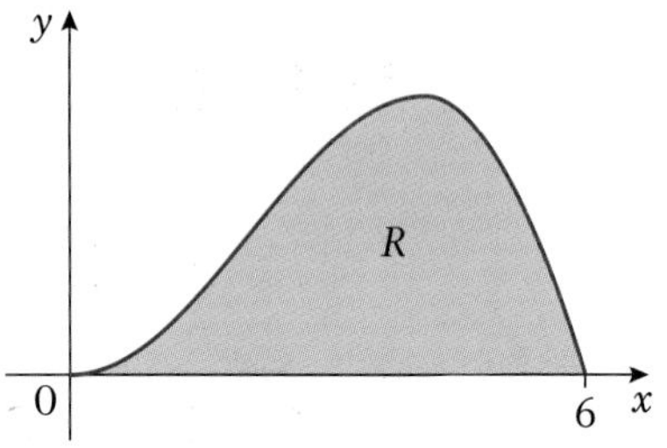

a Show that $y\dfrac{dx}{dt} = 6t^{\frac{1}{2}} - \frac{3}{2}t^{\frac{3}{2}}$.

b Find the area of R.

Summary of key points

1 To find the cartesian equation of a curve given parametrically you eliminate the parameter t between the parametric equations.

2 To find the area under a curve given parametrically you use $\int y\dfrac{dx}{dt}\,dt$.

3 The binomial expansion

In this chapter you will learn and apply the expansion of $(a + x)^n$ for any rational value of n.

3.1 The binomial expansion is

$$(1 + x)^n = 1 + nx + n(n-1)\frac{x^2}{2!} + n(n-1)(n-2)\frac{x^3}{3!} + \dots + {}^nC_r x^r$$

When n is a positive integer, this expansion is finite and exact. This is not generally the case when n is negative or a fraction.

Example 1

Use the binomial expansion to find **a** $(1 + x)^4$ **b** $(1 - 2x)^3$

a $(1 + x)^4$

$= 1 + 4x + \frac{4 \times 3x^2}{2!}$ — Replace n by 4 in the formula.

$+ \frac{4 \times 3 \times 2x^3}{3!}$

$+ \frac{4 \times 3 \times 2 \times 1x^4}{4!}$

$+ \frac{4 \times 3 \times 2 \times 1 \times 0x^5}{5!}$

$= 1 + 4x + \frac{4 \times 3x^2}{2}$

$+ \frac{4 \times 3 \times 2x^3}{6}$

$+ \frac{4 \times 3 \times 2 \times 1x^4}{24}$

$+ \frac{4 \times 3 \times 2 \times 1 \times 0x^5}{120}$ — Simplify coefficients.

$= 1 + 4x + 6x^2 + 4x^3 + 1x^4 + 0x^5$ — All terms after this will also have zero as a coefficient.

$= 1 + 4x + 6x^2 + 4x^3 + x^4$

b $(1 - 2x)^2$

$$= 1 + 3 \times (-2x)$$

Replace n by 3 and x by $-2x$.

$$+ \frac{3 \times 2 \times (-2x)^2}{2!}$$

$$+ \frac{3 \times 2 \times 1 \times (-2x)^3}{3!}$$

$$+ \frac{3 \times 2 \times 1 \times 0 \times (-2x)^4}{4!}$$

$$= 1 - 6x$$

$$+ \frac{3 \times 2 \times 4x^2}{2}$$

$$+ \frac{3 \times 2 \times 1 \times -8x^3}{6}$$

$$+ \frac{3 \times 2 \times 1 \times 0 \times 16x^4}{24}$$

Simplify coefficients.

$$= 1 - 6x + 12x^2 - 8x^3 + 0x^4$$

All terms after this will also have zero as a coefficient.

$$= 1 - 6x + 12x^2 - 8x^3$$

Example 2

Use the binomial expansion to find the first four terms of **a** $\dfrac{1}{(1 + x)}$ **b** $\sqrt{(1 - 3x)}$

a $\dfrac{1}{1 + x} = (1 + x)^{-1}$

Write in index form.

$$= 1 + (-1)(x)$$

Replace n by -1 in the expansion.

$$+ \frac{(-1)(-2)(x)^2}{2!}$$

$$+ \frac{(-1)(-2)(-3)(x)^3}{3!} + \ldots$$

$$= 1 - 1x + 1x^2 - 1x^3 + \ldots$$

$$= 1 - x + x^2 - x^3 + \ldots$$

As n is not a positive integer, no coefficient will ever be equal to zero.
The expansion is **infinite**, and convergent when $|x| < 1$.

b $\sqrt{(1-3x)} = (1-3x)^{\frac{1}{2}}$ — Write in index form.

$= 1 + (\frac{1}{2})(-3x)$ — Replace n by $\frac{1}{2}$ and x by $-3x$.

$+ \dfrac{(\frac{1}{2})(\frac{1}{2}-1)(-3x)^2}{2!}$ — Be careful to write this as $9x^2$, not $-3x^2$.

$+ \dfrac{(\frac{1}{2})(\frac{1}{2}-1)(\frac{1}{2}-2)(-3x)^3}{3!} + \ldots$

$= 1 - \dfrac{3x}{2} + \dfrac{(\frac{1}{2})(-\frac{1}{2})9x^2}{2}$ — Simplify terms.

$+ \dfrac{(\frac{1}{2})(-\frac{1}{2})(-\frac{3}{2})(-27x^3)}{6} + \ldots$

$= 1 - \dfrac{3x}{2} - \dfrac{9x^2}{8} - \dfrac{27x^3}{16} + \ldots$

Because n is not a positive integer, no coefficient will ever be equal to zero.
The expansion is **infinite**, and convergent when $|x| < \frac{1}{3}$ because $|3x| < 1$

Example 3

Find the binomial expansions of **a** $(1-x)^{\frac{1}{3}}$, **b** $\dfrac{1}{(1+4x)^2}$, up to and including the term in x^3.

State the range of values of x for which the expansions are valid.

a $(1-x)^{\frac{1}{3}}$

$= 1 + (\frac{1}{3})(-x)$ — Replace n by $\frac{1}{3}$, x by $(-x)$.

$+ \dfrac{(\frac{1}{3})(\frac{1}{3}-1)(-x)^2}{2!}$ — Simplify brackets.

$+ \dfrac{(\frac{1}{3})(\frac{1}{3}-1)(\frac{1}{3}-2)(-x)^3}{3!} + \ldots$

$= 1 + (\frac{1}{3})(-x) + \dfrac{(\frac{1}{3})(-\frac{2}{3})(-x)^2}{2}$ — Simplify coefficients.

$+ \dfrac{(\frac{1}{3})(-\frac{2}{3})(-\frac{5}{3})(-x)^3}{6} + \ldots$

$= 1 - \frac{1}{3}x - \frac{1}{9}x^2 - \frac{5}{81}x^3 + \ldots$

Expansion is valid as long as $|-x| < 1$ — Terms in expansion are $(-x)$, $(-x)^2$, $(-x)^3$.

$\Rightarrow |x| < 1$

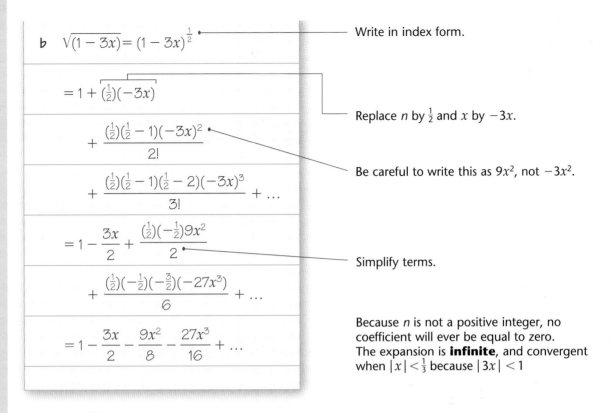

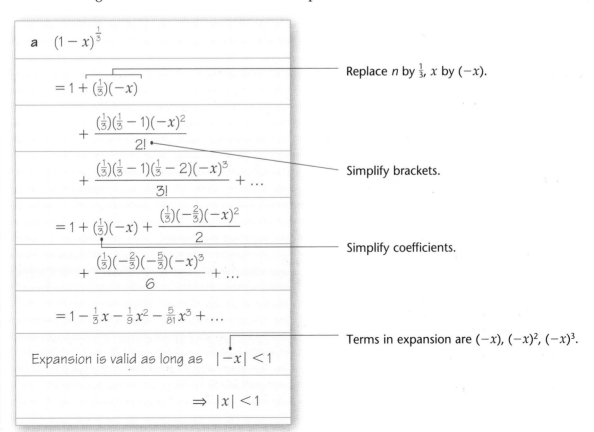

b $\frac{1}{(1+4x)^2}$ — Write in index form.

$= (1+4x)^{-2}$ — Replace n by -2, x by '$4x$'.

$= 1 + (-2)(4x) + \frac{(-2)(-2-1)(4x)^2}{2!} + \frac{(-2)(-2-1)(-2-2)(4x)^3}{3!} + \ldots$ — Simplify brackets.

$= 1 + (-2)(4x) + \frac{(-2)(-3)16x^2}{2} + \frac{(-2)(-3)(-4)64x^3}{6} + \ldots$ — Simplify coefficients.

$= 1 - 8x + 48x^2 - 256x^3 + \ldots$ — Terms in expansion are $(4x)$, $(4x)^2$, $(4x)^3$.

Expansion is valid as long as $|4x| < 1$

$\Rightarrow |x| < \frac{1}{4}$.

Example 4

Find the expansion of $\sqrt{(1-2x)}$ up to and including the term in x^3. By substituting in $x = 0.01$, find a suitable decimal approximation to $\sqrt{2}$.

$\sqrt{(1-2x)} = (1-2x)^{\frac{1}{2}}$ — Write in index form.

Replace n by $\frac{1}{2}$, x by $(-2x)$.

$= 1 + (\frac{1}{2})(-2x) + \frac{(\frac{1}{2})(\frac{1}{2}-1)(-2x)^2}{2!} + \frac{(\frac{1}{2})(\frac{1}{2}-1)(\frac{1}{2}-2)(-2x)^3}{3!} + \ldots$ — Simplify brackets.

$= 1 + (\frac{1}{2})(-2x) + \frac{(\frac{1}{2})(-\frac{1}{2})(4x^2)}{2} + \frac{(\frac{1}{2})(-\frac{1}{2})(-\frac{3}{2})(-8x^3)}{6} + \ldots$ — Simplify coefficients.

$= 1 - x - \frac{x^2}{2} - \frac{x^3}{2} + \ldots$ — Terms in expansion are $(-2x)$, $(-2x)^2$, $(-2x)^3$.

Expansion is valid if $|2x| < 1$

$\Rightarrow |x| < \frac{1}{2}$.

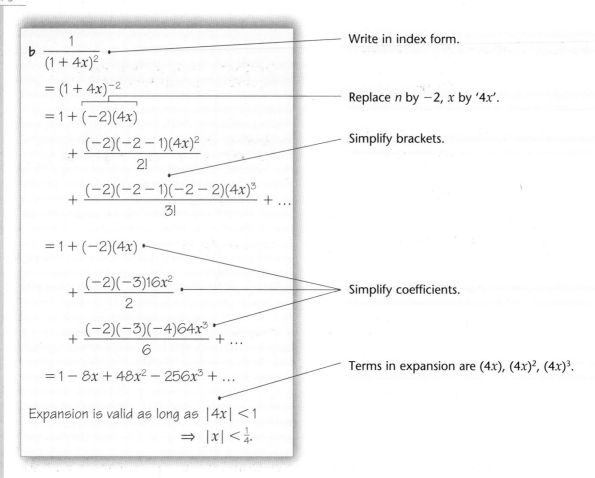

$\sqrt{(1 - 2 \times 0.01)} \approx 1 - 0.01 - \dfrac{(0.01)^2}{2} - \dfrac{(0.01)^3}{2}$	Substitute $x = 0.01$ into both sides of expansion. This is valid as $\lvert x \rvert < \frac{1}{2}$.
$\sqrt{0.98} \approx 1 - 0.01 - 0.00005 - 0.0000005$	Simplify both sides. Note that the terms are getting smaller.
$\sqrt{\dfrac{98}{100}} \approx 0.9899495$	Write 0.98 as $\frac{98}{100}$.
$\sqrt{\dfrac{49 \times 2}{100}} \approx 0.9899495$	Use rules of surds.
$\dfrac{7\sqrt{2}}{10} \approx 0.9899495$	$\times 10$, $\div 7$
$\sqrt{2} \approx \dfrac{0.9899495 \times 10}{7}$	Simplify.
$\sqrt{2} \approx 1.414213571$	

Exercise 3A

1 Find the binomial expansion of the following up to and including the terms in x^3. State the range values of x for which these expansions are valid.

a $(1 + 2x)^3$ **b** $\dfrac{1}{1 - x}$ **c** $\sqrt{(1 + x)}$ **d** $\dfrac{1}{(1 + 2x)^3}$

e $\sqrt[3]{(1 - 3x)}$ **f** $(1 - 10x)^{\frac{3}{2}}$ **g** $\left(1 + \dfrac{x}{4}\right)^{-4}$ **h** $\dfrac{1}{(1 + 2x^2)}$

2 By first writing $\dfrac{(1 + x)}{(1 - 2x)}$ as $(1 + x)(1 - 2x)^{-1}$ show that the cubic approximation to $\dfrac{(1 + x)}{(1 - 2x)}$ is $1 + 3x + 6x^2 + 12x^3$. State the range of values of x for which this expansion is valid.

3 Find the binomial expansion of $\sqrt{(1 + 3x)}$ in ascending powers of x up to and including the term in x^3. By substituting $x = 0.01$ in the expansion, find an approximation to $\sqrt{103}$. By comparing it with the exact value, comment on the accuracy of your approximation.

4 In the expansion of $(1 + ax)^{-\frac{1}{2}}$ the coefficient of x^2 is 24. Find possible values of the constant a and the corresponding term in x^3.

5 Show that if x is small, the expression $\sqrt{\left(\dfrac{1 + x}{1 - x}\right)}$ is approximated by $1 + x + \frac{1}{2}x^2$.

6 Find the first four terms in the expansion of $(1 - 3x)^{\frac{3}{2}}$. By substituting in a suitable value of x, find an approximation to $97^{\frac{3}{2}}$.

3.2 You can use the binomial expansion of $(1 + x)^n$ to expand $(a + bx)^n$ for any constants a and b by simply taking out a as a factor.

Example 5

Find the first four terms in the binomial expansion of **a** $\sqrt{(4 + x)}$ **b** $\dfrac{1}{(2 + 3x)^2}$.

State the range in values of x for which these expansions are valid.

a $\sqrt{(4+x)} = (4+x)^{\frac{1}{2}}$ — Write in index form.

$= \left[4\left(1+\dfrac{x}{4}\right)\right]^{\frac{1}{2}}$ — Take out a factor of 4.

$= 4^{\frac{1}{2}}\left(1+\dfrac{x}{4}\right)^{\frac{1}{2}}$

$= 2\left(1+\dfrac{x}{4}\right)^{\frac{1}{2}}$ — Write $4^{\frac{1}{2}}$ as 2.

$$= 2\left[1 + \left(\frac{1}{2}\right)\left(\frac{x}{4}\right) + \frac{\left(\frac{1}{2}\right)\left(\frac{1}{2}-1\right)\left(\frac{x}{4}\right)^2}{2!} + \frac{\left(\frac{1}{2}\right)\left(\frac{1}{2}-1\right)\left(\frac{1}{2}-2\right)\left(\frac{x}{4}\right)^3}{3!} + \ldots\right]$$

Expand $\left(1+\dfrac{x}{4}\right)^{\frac{1}{2}}$ using the binomial expansion with $n = \frac{1}{2}$ and $x = \dfrac{x}{4}$.

$$= 2\left[1 + \left(\frac{1}{2}\right)\left(\frac{x}{4}\right) + \frac{\left(\frac{1}{2}\right)\left(-\frac{1}{2}\right)\left(\frac{x^2}{16}\right)}{2} + \frac{\left(\frac{1}{2}\right)\left(-\frac{1}{2}\right)\left(-\frac{3}{2}\right)\left(\frac{x^3}{64}\right)}{6} + \ldots\right]$$

Simplify coefficients.

$$= 2\left[1 + \frac{x}{8} - \frac{x^2}{128} + \frac{x^3}{1024} + \ldots\right]$$

Multiply by the 2.

$$= 2 + \frac{x}{4} - \frac{x^2}{64} + \frac{x^3}{512} + \ldots$$

Expansion is valid if $\left|\dfrac{x}{4}\right| < 1$

$\Rightarrow |x| < 4.$

Terms in expansion are $\left(\dfrac{x}{4}\right), \left(\dfrac{x}{4}\right)^2, \left(\dfrac{x}{4}\right)^3$.

b $\dfrac{1}{(2+3x)^2} = (2+3x)^{-2}$ — Write in index form.

$= \left[2\left(1+\dfrac{3x}{2}\right)\right]^{-2}$ — Take out a factor of 2.

$= 2^{-2}\left(1+\dfrac{3x}{2}\right)^{-2}$ — Write $2^{-2} = \dfrac{1}{2^2} = \dfrac{1}{4}$

$= \dfrac{1}{4}\left(1+\dfrac{3x}{2}\right)^{-2}$

$= \dfrac{1}{4}\left[1 + (-2)\left(\dfrac{3x}{2}\right) + \dfrac{(-2)(-2-1)\left(\dfrac{3x}{2}\right)^2}{2!} + \dfrac{(-2)(-2-1)(-2-2)\left(\dfrac{3x}{2}\right)^3}{3!} + \ldots\right]$

Expand $\left(1+\dfrac{3x}{2}\right)^{-2}$ using the binomial expansion with $n = -2$ and $x = \dfrac{3x}{2}$.

$= \dfrac{1}{4}\left[1 + (-2)\left(\dfrac{3x}{2}\right) + \dfrac{(-2)(-3)\left(\dfrac{9x^2}{4}\right)}{2} + \dfrac{(-2)(-3)(-4)\left(\dfrac{27x^3}{8}\right)}{6} + \ldots\right]$

Simplify coefficients.

$= \dfrac{1}{4}\left[1 - 3x + \dfrac{27x^2}{4} - \dfrac{27x^3}{2} + \ldots\right]$ — Multiply by the $\frac{1}{4}$.

$= \dfrac{1}{4} - \dfrac{3}{4}x + \dfrac{27x^2}{16} - \dfrac{27x^3}{8} + \ldots$

Expansion is valid if $\left|\dfrac{3x}{2}\right| < 1$

$\Rightarrow \quad |x| < \dfrac{2}{3}.$

Terms in expansion are $\left(\dfrac{3x}{2}\right), \left(\dfrac{3x}{2}\right)^2, \left(\dfrac{3x}{2}\right)^3.$

Exercise 3B

1 Find the binomial expansions of the following in ascending powers of x as far as the term in x^3. State the range of values of x for which the expansions are valid.

a $\sqrt{(4+2x)}$ **b** $\dfrac{1}{2+x}$ **c** $\dfrac{1}{(4-x)^2}$ **d** $\sqrt{(9+x)}$

e $\dfrac{1}{\sqrt{(2+x)}}$ **f** $\dfrac{5}{3+2x}$ **g** $\dfrac{1+x}{2+x}$ **h** $\sqrt{\left(\dfrac{2+x}{1-x}\right)}$

2 Prove that if x is sufficiently small, $\dfrac{3+2x-x^2}{4-x}$ may be approximated by $\frac{3}{4}+\frac{11}{16}x-\frac{5}{64}x^2$.
What does 'sufficiently small' mean in this question?

3 Find the first four terms in the expansion of $\sqrt{(4-x)}$. By substituting $x=\frac{1}{9}$, find a fraction that is an approximation to $\sqrt{35}$. By comparing this to the exact value, state the degree of accuracy of your approximation.

4 The expansion of $(a+bx)^{-2}$ may be approximated by $\frac{1}{4}+\frac{1}{4}x+cx^2$. Find the values of the constants a, b and c.

3.3 You can use partial fractions to simplify the expansions of many more difficult expressions.

Example 6

a Express $\dfrac{4-5x}{(1+x)(2-x)}$ as partial fractions.

b Hence show that the cubic approximation of $\dfrac{4-5x}{(1+x)(2-x)}$ is $2-\dfrac{7x}{2}+\dfrac{11}{4}x^2-\dfrac{25}{8}x^3$.

c State the range of values of x for which the expansion is valid.

a $\dfrac{4-5x}{(1+x)(2-x)} \equiv \dfrac{A}{(1+x)}+\dfrac{B}{(2-x)}$ — The denominators must be $(1+x)$ and $(2-x)$.

$\equiv \dfrac{A(2-x)+B(1+x)}{(1+x)(2-x)}$ — Add the fractions.

$4-5x \equiv A(2-x)+B(1+x)$ — Set the numerators equal.

Substitute $x=2$

$4-10 = A\times 0+B\times 3$ — Set $x=2$ to find B.

$-6=3B$

$B=-2$

Substitute $x=-1$

$4+5=A\times 3+B\times 0$ — Set $x=-1$ to find A.

$9=3A$

$A=3$

so $\dfrac{4-5x}{(1+x)(2-x)} = \dfrac{3}{1+x}-\dfrac{2}{2-x}$

b $\dfrac{4-5x}{(1+x)(2-x)} = \dfrac{3}{(1+x)} - \dfrac{2}{(2-x)}$ — Write in index form.

$= 3(1+x)^{-1} - 2(2-x)^{-1}$

The expansion of $3(1+x)^{-1}$

$$= 3\left[1 + (-1)(x) + (-1)(-2)\frac{(x)^2}{2!} + (-1)(-2)(-3)\frac{(x)^3}{3!} + \ldots\right]$$

Expand $3(1 + x)^{-1}$ using the binomial expansion with $n = -1$.

$= 3[1 - x + x^2 - x^3 + \ldots]$

$= 3 - 3x + 3x^2 - 3x^3 + \ldots$

Take out a factor of 2.

The expansion of $2(2-x)^{-1}$

$$= 2\left[2\left(1 - \frac{x}{2}\right)\right]^{-1}$$

$$= 2 \times 2^{-1}\left(1 - \frac{x}{2}\right)^{-1}$$

$$= 1 \times \left[1 + (-1)\left(-\frac{x}{2}\right) + \frac{(-1)(-2)\left(-\frac{x}{2}\right)^2}{2!} + \frac{(-1)(-2)(-3)\left(-\frac{x}{2}\right)^3}{3!} + \ldots\right]$$

Expand $\left(1 - \frac{x}{2}\right)^{-1}$ using the binomial expansion with $n = -1$ and $x = \frac{x}{2}$.

$$= 1 \times \left[1 + \frac{x}{2} + \frac{x^2}{4} + \frac{x^3}{8} + \ldots\right]$$

$$= 1 + \frac{x}{2} + \frac{x^2}{4} + \frac{x^3}{8}$$

Hence $\dfrac{4-5x}{(1+x)(2-x)}$

$= 3(1+x)^{-1} - 2(2-x)^{-1}$ — 'Add' both expressions.

$= (3 - 3x + 3x^2 - 3x^3)$

$- \left(1 + \dfrac{x}{2} + \dfrac{x^2}{4} + \dfrac{x^3}{8}\right)$

$= 2 - \dfrac{7}{2}x + \dfrac{11}{4}x^2 - \dfrac{25}{8}x^3$

c $\dfrac{3}{(1+x)}$ is valid if $|x| < 1$ — Terms are x, x^2, x^3.

$\dfrac{2}{(2-x)}$ is valid if $\left|\dfrac{x}{2}\right| < 1 \Rightarrow |x| < 2$ — Terms are $\dfrac{x}{2}, \left(\dfrac{x}{2}\right)^2, \left(\dfrac{x}{2}\right)^3$.

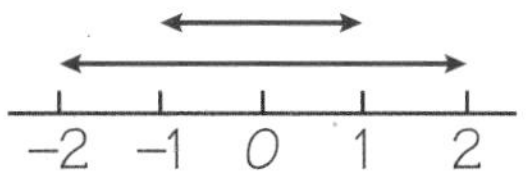

Look for values of x that satisfy both expressions.

Both expressions are valid provided $|x| < 1$.

Exercise 3C

1 **a** Express $\dfrac{8x+4}{(1-x)(2+x)}$ as partial fractions.

b Hence or otherwise expand $\dfrac{8x+4}{(1-x)(2+x)}$ in ascending powers of x as far as the term in x^2.

c State the set of values of x for which the expansion is valid.

2 **a** Express $\dfrac{-2x}{(2+x)^2}$ as a partial fraction.

b Hence prove that $\dfrac{-2x}{(2+x)^2}$ can be expressed in the form $0 - \dfrac{1}{2}x + Bx^2 + Cx^3$ where constants B and C are to be determined.

c State the set of values of x for which the expansion is valid.

3 **a** Express $\dfrac{6+7x+5x^2}{(1+x)(1-x)(2+x)}$ as a partial fraction.

b Hence or otherwise expand $\dfrac{6+7x+5x^2}{(1+x)(1-x)(2+x)}$ in ascending powers of x as far as the term in x^3.

c State the set of values of x for which the expansion is valid.

Mixed exercise 3D

1 Find binomial expansions of the following in ascending powers of x as far as the term in x^3. State the set of values of x for which the expansion is valid.

a $(1 - 4x)^3$ **b** $\sqrt{(16 + x)}$ **c** $\dfrac{1}{(1 - 2x)}$ **d** $\dfrac{4}{2 + 3x}$

e $\dfrac{4}{\sqrt{(4 - x)}}$ **f** $\dfrac{1 + x}{1 + 3x}$ **g** $\left(\dfrac{1 + x}{1 - x}\right)^2$ **h** $\dfrac{x - 3}{(1 - x)(1 - 2x)}$

2 Find the first four terms of the expansion in ascending powers of x of:

$(1 - \frac{1}{2}x)^{\frac{1}{2}}$, $|x| < 2$

and simplify each coefficient. **E**

(adapted)

3 Show that if x is sufficiently small then $\dfrac{3}{\sqrt{(4 + x)}}$ can be approximated by $\dfrac{3}{2} - \dfrac{3}{16}x + \dfrac{9}{256}x^2$.

4 Given that $|x| < 4$, find, in ascending powers of x up to and including the term in x^3, the series expansion of:

a $(4 - x)^{\frac{1}{2}}$ **b** $(4 - x)^{\frac{1}{2}}(1 + 2x)$ **E**

(adapted)

5 **a** Find the first four terms of the expansion, in ascending powers of x, of $(2 + 3x)^{-1}$, $|x| < \frac{2}{3}$

b Hence or otherwise, find the first four non-zero terms of the expansion, in ascending powers of x, of:

$\dfrac{1 + x}{2 + 3x}$, $|x| < \frac{2}{3}$ **E**

6 Find, in ascending powers of x up to and including the term in x^3, the series expansion of $(4 + x)^{-\frac{1}{2}}$, giving your coefficients in their simplest form. **E**

7 $f(x) = (1 + 3x)^{-1}$, $|x| < \frac{1}{3}$.

a Expand $f(x)$ in ascending powers of x up to and including the term in x^3.

b Hence show that, for small x:

$$\frac{1 + x}{1 + 3x} \approx 1 - 2x + 6x^2 - 18x^3.$$

c Taking a suitable value for x, which should be stated, use the series expansion in part **b** to find an approximate value for $\frac{101}{103}$, giving your answer to 5 decimal places. **E**

8 Obtain the first four non-zero terms in the expansion, in ascending powers of x, of the function $f(x)$ where $f(x) = \dfrac{1}{\sqrt{(1 + 3x^2)}}$, $3x^2 < 1$. **E**

9 Give the binomial expansion of $(1 + x)^{\frac{1}{2}}$ up to and including the term in x^3. By substituting $x = \frac{1}{4}$, find the fraction that is an approximation to $\sqrt{5}$.

10 When $(1 + ax)^n$ is expanded as a series in ascending powers of x, the coefficients of x and x^2 are -6 and 27 respectively.

a Find the values of a and n.

b Find the coefficient of x^3.

c State the values of x for which the expansion is valid. **E** (adapted)

11 **a** Express $\dfrac{9x^2 + 26x + 20}{(1 + x)(2 + x)^2}$ as a partial fraction.

b Hence or otherwise show that the expansion of $\dfrac{9x^2 + 26x + 20}{(1 + x)(2 + x)^2}$ in ascending powers of x can be approximated to $5 - \dfrac{7x}{2} + Bx^2 + Cx^3$ where B and C are constants to be found.

c State the set of values of x for which this expansion is valid.

Summary of key points

1 The binomial expansion $(1 + x)^n = 1 + nx + \dfrac{n(n-1)x^2}{2!} + \dfrac{n(n-1)(n-2)x^3}{3!} + \ldots$ can be used to give an exact expression if n is a positive integer, or an approximate expression for any other rational number.

- $(1 + 2x)^3 = 1 + 3(2x) + 3 \times 2\,\dfrac{(2x)^2}{2!} + 3 \times 2 \times 1 \times \dfrac{(2x)^3}{3!} + 3 \times 2 \times 1 \times 0 \times \dfrac{(2x)^4}{4!}$

  $= 1 + 6x + 12x^2 + 8x^3$ (Expansion is *finite* and *exact.*)

- $\sqrt{(1 - x)} = (1 - x)^{\frac{1}{2}} = 1 + \left(\dfrac{1}{2}\right)(-x) + \left(\dfrac{1}{2}\right)\left(-\dfrac{1}{2}\right)\dfrac{(-x)^2}{2!} + \left(\dfrac{1}{2}\right)\left(-\dfrac{1}{2}\right)\left(-\dfrac{3}{2}\right)\dfrac{(-x)^3}{3!} + \ldots$

  $= 1 - \dfrac{1}{2}x - \dfrac{1}{8}x^2 - \dfrac{1}{16}x^3 + \ldots$

  (Expansion is *infinite* and *approximate.*)

2 The expansion $(1 + x)^n = 1 + nx + n(n-1)\dfrac{x^2}{2!} + n(n-1)(n-2)\dfrac{x^3}{3!} + \ldots$, where n is negative or a fraction, is only valid if $|x| < 1$.

3 You can adapt the binomial expansion to include expressions of the form $(a + bx)^n$ by simply taking out a common factor of a:

e.g. $\dfrac{1}{(3 + 4x)} = (3 + 4x)^{-1} = \left[3\left(1 + \dfrac{4x}{3}\right)\right]^{-1}$

$= 3^{-1}\left(1 + \dfrac{4x}{3}\right)^{-1}$

4 You can use knowledge of partial fractions to expand more difficult expressions, e.g.

$\dfrac{7 + x}{(3 - x)(2 + x)} = \dfrac{2}{(3 - x)} + \dfrac{1}{(2 + x)}$

$= 2(3 - x)^{-1} + (2 + x)^{-1}$

$= \dfrac{2}{3}\left(1 - \dfrac{x}{3}\right)^{-1} + \dfrac{1}{2}\left(1 + \dfrac{x}{2}\right)^{-1}$

4 Differentiation

In this chapter you will learn to: find the gradient of a curve which is given in parametric form; differentiate relations given implicitly; find the rate of change of exponential and decay functions; connect rates of change of related variables; set up differential equations.

4.1 You can find the gradient of a curve given in parametric coordinates.

When a curve is described by parametric equations

- You differentiate x and y with respect to the parameter t.
- Then you use the chain rule rearranged into the form $\frac{dy}{dx} = \frac{dy}{dt} \div \frac{dx}{dt}$.

Example 1

Find the gradient at the point P where $t = 2$, on the curve given parametrically by $x = t^3 + t$, $y = t^2 + 1$, $t \in \mathbb{R}$.

$\frac{dx}{dt} = 3t^2 + 1, \frac{dy}{dt} = 2t$

First differentiate x and y with respect to the parameter t.

$$\frac{dy}{dx} = \frac{\frac{dy}{dt}}{\frac{dx}{dt}} = \frac{2t}{3t^2 + 1}$$

Use the chain rule $\frac{dy}{dx} \times \frac{dx}{dt} = \frac{dy}{dt}$ and rearrange to give $\frac{dy}{dx}$.

When $t = 2$, $\frac{dy}{dx} = \frac{4}{13}$

Substitute $t = 2$ into $\frac{2t}{3t^2 + 1}$.

So the gradient at P is $\frac{4}{13}$.

Example 2

Find the equation of the normal at the point P where $\theta = \frac{\pi}{6}$, on the curve with parametric equations $x = 3 \sin \theta$, $y = 5 \cos \theta$.

$\frac{dx}{d\theta} = 3 \cos \theta, \frac{dy}{d\theta} = -5 \sin \theta$

First differentiate x and y with respect to the parameter θ.

$$\therefore \frac{dy}{dx} = \frac{-5 \sin \theta}{3 \cos \theta}$$

At point P, where $\theta = \frac{\pi}{6}$,

$$\frac{dy}{dx} = \frac{-5 \times \frac{1}{2}}{3 \times \frac{\sqrt{3}}{2}} = \frac{-5}{3\sqrt{3}}$$

Use the chain rule, $\frac{dy}{d\theta} \div \frac{dx}{d\theta}$, and substitute $\theta = \frac{\pi}{6}$.

The gradient of the normal at P

is $\frac{3\sqrt{3}}{5}$, and at P, $x = \frac{3}{2}$, $y = \frac{5\sqrt{3}}{2}$.

The normal is perpendicular to the curve, so its gradient is $-\frac{1}{m}$, where m is the gradient of the curve at that point.

The equation of the normal is

$$y - \frac{5\sqrt{3}}{2} = \frac{3\sqrt{3}}{5}\left(x - \frac{3}{2}\right)$$

Use equation for a line in the form $(y - y_1) = m(x - x_1)$.

$$\therefore \quad 5y = 3\sqrt{3}x + 8\sqrt{3}$$

Exercise 4A

1 Find $\frac{dy}{dx}$ for each of the following, leaving your answer in terms of the parameter t:

a $x = 2t$, $y = t^2 - 3t + 2$ **b** $x = 3t^2$, $y = 2t^3$ **c** $x = t + 3t^2$, $y = 4t$

d $x = t^2 - 2$, $y = 3t^5$ **e** $x = \frac{2}{t}$, $y = 3t^2 - 2$ **f** $x = \frac{1}{2t-1}$, $y = \frac{t^2}{2t-1}$

g $x = \frac{2t}{1+t^2}$, $y = \frac{1-t^2}{1+t^2}$ **h** $x = t^2e^t$, $y = 2t$ **i** $x = 4\sin 3t$, $y = 3\cos 3t$

j $x = 2 + \sin t$, $y = 3 - 4\cos t$ **k** $x = \sec t$, $y = \tan t$ **l** $x = 2t - \sin 2t$, $y = 1 - \cos 2t$

2 **a** Find the equation of the tangent to the curve with parametric equations $x = 3t - 2\sin t$, $y = t^2 + t\cos t$, at the point P, where $t = \frac{\pi}{2}$.

b Find the equation of the tangent to the curve with parametric equations $x = 9 - t^2$, $y = t^2 + 6t$, at the point P, where $t = 2$.

3 **a** Find the equation of the normal to the curve with parametric equations $x = e^t$, $y = e^t + e^{-t}$, at the point P, where $t = 0$.

b Find the equation of the normal to the curve with parametric equations $x = 1 - \cos 2t$, $y = \sin 2t$, at the point P, where $t = \frac{\pi}{6}$.

4 Find the points of zero gradient on the curve with parametric equations

$$x = \frac{t}{1-t},\ y = \frac{t^2}{1-t},\ t \neq 1.$$

You do not need to establish whether they are maximum or minimum points.

4.2 You can differentiate relations which are implicit, such as $x^2 + y^2 = 8x$, and $\cos(x + y) = \sin y$.

Differentiate each term in turn using the chain rule and the product rule, as appropriate:

- $\frac{\mathrm{d}}{\mathrm{d}x}(y^n) = ny^{n-1}\frac{\mathrm{d}y}{\mathrm{d}x}$ — By the chain rule.
- $\frac{\mathrm{d}}{\mathrm{d}x}(xy) = x\frac{\mathrm{d}}{\mathrm{d}x}(y) + y\frac{\mathrm{d}}{\mathrm{d}x}(x)$

$= x\frac{\mathrm{d}y}{\mathrm{d}x} + y \times 1$ — By the product rule.

$= x\frac{\mathrm{d}y}{\mathrm{d}x} + y$

Example 3

Find $\frac{\mathrm{d}y}{\mathrm{d}x}$ in terms of x and y where $x^3 + x + y^3 + 3y = 6$.

$$3x^2 + 1 + 3y^2\frac{dy}{dx} + 3\frac{dy}{dx} = 0$$

Differentiate the expression term by term with respect to x.

Use the chain rule to differentiate y^3.

$$\frac{dy}{dx}(3y^2 + 3) = -3x^2 - 1$$

Then make $\frac{\mathrm{d}y}{\mathrm{d}x}$ the subject of the formula.

$$\frac{dy}{dx} = -\frac{(3x^2 + 1)}{3(1 + y^2)}$$

Divide by the coefficient of $\frac{\mathrm{d}y}{\mathrm{d}x}$ and factorise.

Example 4

Find the value of $\frac{\mathrm{d}y}{\mathrm{d}x}$ at the point (1, 1)

where $4xy^2 + \frac{6x^2}{y} = 10$.

$$\left(4x \times 2y\frac{dy}{dx} + 4y^2\right) + \frac{12x}{y} - \frac{6x^2}{y^2}\frac{dy}{dx} = 0$$

Differentiate each term with respect to x.

Use the product rule on each term, expressing $\frac{6x^2}{y}$ as $6x^2y^{-1}$.

Substitute $x = 1$, $y = 1$ to give

Find the value of $\frac{\mathrm{d}y}{\mathrm{d}x}$ at (1, 1) by substituting $x = 1$, $y = 1$.

$$\left(8\frac{dy}{dx} + 4\right) + 12 - 6\frac{dy}{dx} = 0$$

Substitute before rearranging, as this simplifies the working.

i.e. $16 + 2\frac{dy}{dx} = 0$

$$\frac{dy}{dx} = -8$$

Finally make $\frac{\mathrm{d}y}{\mathrm{d}x}$ the subject of the formula.

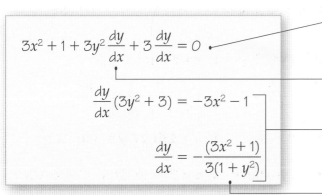

Example 5

Find the value of $\frac{dy}{dx}$ at the point (1, 1) where $e^{2x}\ln y = x + y - 2$.

$$e^{2x} \times \frac{1}{y}\frac{dy}{dx} + \ln y \times 2e^{2x} = 1 + \frac{dy}{dx}$$

Differentiate each term with respect to x.

Use the product rule applied to the term on the left hand side of the equation, noting that $\ln y$ differentiates to give $\frac{1}{y}\frac{dy}{dx}$.

Substitute $x = 1$, $y = 1$ to give

$$e^2 \times \frac{dy}{dx} = 1 + \frac{dy}{dx}$$

$$\therefore \quad (e^2 - 1)\frac{dy}{dx} = 1$$

$$\frac{dy}{dx} = \frac{1}{e^2 - 1}$$

Rearrange to make $\frac{dy}{dx}$ the subject of the formula.

■ **In an implicit equation:**

- **Note that when f(y) is differentiated with respect to x it becomes $f'(y)\frac{dy}{dx}$.**
- **A product term such as f(x).g(y) is differentiated by the product rule and becomes $f(x).g'(y)\frac{dy}{dx} + g(y).f'(x)$.**

Exercise 4B

1 Find an expression in terms of x and y for $\frac{dy}{dx}$, given that:

a $x^2 + y^3 = 2$ **b** $x^2 + 5y^2 = 14$ **c** $x^2 + 6x - 8y + 5y^2 = 13$

d $y^3 + 3x^2y - 4x = 0$ **e** $3y^2 - 2y + 2xy = x^3$ **f** $x = \frac{2y}{x^2 - y}$

g $(x - y)^4 = x + y + 5$ **h** $e^x y = xe^y$ **i** $\sqrt{(xy)} + x + y^2 = 0$

2 Find the equation of the tangent to the curve with implicit equation $x^2 + 3xy^2 - y^3 = 9$ at the point (2, 1).

3 Find the equation of the normal to the curve with implicit equation $(x + y)^3 = x^2 + y$ at the point (1, 0).

4 Find the coordinates of the points of zero gradient on the curve with implicit equation $x^2 + 4y^2 - 6x - 16y + 21 = 0$.

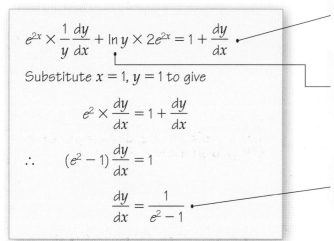

4.3 You can differentiate the general power function a^x, where a is constant.

This function describes growth and decay and its derivative gives a measure of the rate of change of this growth or decay.

Example 6

Differentiate $y = a^x$, where a is a constant.

Since $y = a^x$

$\ln y = \ln a^x$

$\therefore \quad \ln y = x \ln a$

Take logs of both sides, then use properties of logs to express $\ln a^x$ as $x \ln a$.

$\therefore \quad \frac{1}{y}\frac{dy}{dx} = \ln a$

Use implicit differentiation to differentiate $\ln y$.

$\therefore \quad \frac{dy}{dx} = y \ln a$

$= a^x \ln a$

Replace y by a^x.

■ **If $y = a^x$, then $\frac{dy}{dx} = a^x \ln a$**

You should learn this result.

(In particular, if $y = e^x$, then $\frac{dy}{dx} = e^x \ln e = e^x$, as you know from the C3 book.)

Exercise 4C

1 Find $\frac{dy}{dx}$ for each of the following:

a $y = 3^x$ **b** $y = \left(\frac{1}{2}\right)^x$ c $y = xa^x$ **d** $y = \frac{2^x}{x}$

2 Find the equation of the tangent to the curve $y = 2^x + 2^{-x}$ at the point $(2, 4\frac{1}{4})$.

3 A particular radioactive isotope has an activity R millicuries at time t days given by the equation $R = 200(0.9)^t$. Find the value of $\frac{dR}{dt}$, when $t = 8$.

4 The population of Cambridge was 37 000 in 1900 and was about 109 000 in 2000. Find an equation of the form $P = P_0k^t$ to model this data, where t is measured as years since 1900. Evaluate $\frac{dP}{dt}$ in the year 2000. What does this value represent?

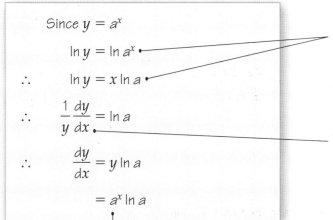

4.4 You can relate one rate of change to another.

You can use the chain rule once, or several times, to connect the rates of change in a question involving more than two variables.

Example 7

Given that the area of a circle $A\,\text{cm}^2$ is related to its radius $r\,\text{cm}$ by the formula $A = \pi r^2$, and that the rate of change of its radius in cm s^{-1} is given by $\frac{\mathrm{d}r}{\mathrm{d}t} = 5$, find $\frac{\mathrm{d}A}{\mathrm{d}t}$ when $r = 3$.

$A = \pi r^2$

$\therefore \quad \frac{dA}{dr} = 2\pi r$

As A is a function of r, find $\frac{dA}{dr}$.

Using $\frac{dA}{dt} = \frac{dA}{dr} \times \frac{dr}{dt}$

You should use the chain rule, giving the derivative which you need to find in terms of known derivatives.

$\frac{dA}{dt} = 2\pi r \times 5$

$= 30\pi$, when $r = 3$.

Example 8

The volume of a hemisphere $V\,\text{cm}^3$ is related to its radius $r\,\text{cm}$ by the formula $V = \frac{2}{3}\pi r^3$ and the total surface area $S\,\text{cm}^2$ is given by the formula $S = \pi r^2 + 2\pi r^2 = 3\pi r^2$. Given that the rate of increase of volume, in $\text{cm}^3\,\text{s}^{-1}$, $\frac{\mathrm{d}V}{\mathrm{d}t} = 6$, find the rate of increase of surface area area $\frac{\mathrm{d}S}{\mathrm{d}t}$.

$V = \frac{2}{3}\pi r^3$ and $S = 3\pi r^2$

This is area of circular base plus area of curved surface.

$\frac{dV}{dr} = 2\pi r^2$ and $\frac{dS}{dr} = 6\pi r$

As V and S are functions of r, find $\frac{dV}{dr}$ and $\frac{dS}{dr}$.

Now $\frac{dS}{dt} = \frac{dS}{dr} \times \frac{dr}{dV} \times \frac{dV}{dt}$

$= 6\pi r \times \frac{1}{2\pi r^2} \times 6$

Use an extended chain rule together with the property that $\frac{dr}{dV} = 1 \div \frac{dV}{dr}$.

$= \frac{18}{r}$

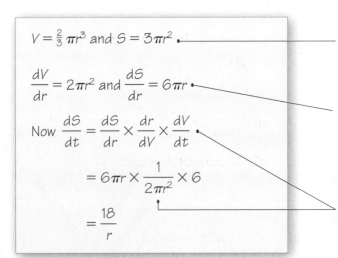

Exercise 4D

1 Given that $V = \frac{1}{3}\pi r^3$ and that $\frac{dV}{dt} = 8$, find $\frac{dr}{dt}$ when $r = 3$.

2 Given that $A = \frac{1}{4}\pi r^2$ and that $\frac{dr}{dt} = 6$, find $\frac{dA}{dt}$ when $r = 2$.

3 Given that $y = xe^x$ and that $\frac{dx}{dt} = 5$, find $\frac{dy}{dt}$ when $x = 2$.

4 Given that $r = 1 + 3\cos\theta$ and that $\frac{d\theta}{dt} = 3$, find $\frac{dr}{dt}$ when $\theta = \frac{\pi}{6}$.

4.5 You can set up a differential equation from information given in context.

Differential equations arise from many problems in mechanics, physics, chemistry, biology and economics. As their name suggests, these equations involve differential coefficients and so equations of the form

$$\frac{dy}{dx} = 3y, \quad \frac{ds}{dt} = 2 + 6t, \quad \frac{d^2y}{dt^2} = -25y \quad \frac{dP}{dt} = 10 - 4P$$

are differential equations (x, y, t, s and P are variables).

In the C4 book you will consider only first order differential equations, which involve first derivatives only.

■ **You can set up simple differential equations from information given in context. This may involve using connected rates of change, or ideas of proportion.**

Example 9

In the decay of radioactive particles the rate at which particles decay is proportional to the number of particles remaining. Write down an equation for the rate of change of the number of particles.

Let N be the number of particles and let t be time. The rate of change of the number of particles $\frac{dN}{dt}$ decays at a rate proportional to N.

Note that this is a proportional problem.

i.e. $\frac{dN}{dt} = -kN$, where k is a positive constant.

$\frac{dN}{dt} \propto N \rightarrow \frac{dN}{dt} = cN$

c is the constant of proportion.

The minus sign arises because the number of particles is decreasing.

Example 10

A population is growing at a rate which is proportional to the size of the population at a given time. Write down an equation for the rate of growth of the population.

Let P be the population and t be the time.

The rate of change of the population $\frac{dP}{dt}$ grows at a rate proportional to P.

i.e. $\frac{dP}{dt} = kP$, where k is a positive constant.

The population is increasing, so there is no minus sign.

k is the constant of proportion.

Example 11

Newton's Law of Cooling states that the rate of loss of temperature of a body is proportional to the excess temperature of the body over its surroundings. Write an equation that expresses this law.

Let the temperature of the body be θ degrees and the time be t seconds.

The rate of change of the temperature $\frac{d\theta}{dt}$ decreases at a rate proportional to $(\theta - \theta_0)$, where θ_0 is the temperature of the surroundings.

i.e. $\frac{d\theta}{dt} = -k(\theta - \theta_0)$, where k is a positive constant.

$(\theta - \theta_0)$ is the difference between the temperature of the body and that of its surroundings.

The minus sign arises because the temperature is decreasing. The question mentions loss of temperature.

Example 12

The head of a snowman of radius R cm loses volume by evaporation at a rate proportional to its surface area. Assuming that the head is spherical, that the volume of a sphere is $\frac{4}{3}\pi R^3$ cm^3 and that the surface is $4\pi R^2$ cm^2, write down a differential equation for the rate of change of radius of the snowman's head.

The first sentence tells you that $\frac{dV}{dt} = -kA$, where V cm^3 is the volume, t seconds is time, k is a positive constant and A cm^2 is the surface area referred to in the question.

Since $V = \frac{4}{3}\pi R^3$

$$\frac{dV}{dR} = 4\pi R^2$$

$$\therefore \quad \frac{dV}{dt} = \frac{dV}{dR} \times \frac{dR}{dt} = 4\pi R^2 \times \frac{dR}{dt}$$

But as $\frac{dV}{dt} = -kA$

$$4\pi R^2 \times \frac{dR}{dt} = -k \times 4\pi R^2$$

$$\therefore \quad \frac{dR}{dt} = -k$$

The question asks for a differential equation in terms of R, and so you need to use the expressions for V and A in terms of R.

The chain rule is used here because this is a related rate of change.

Divide both sides by the common factor $4\pi R^2$.

This gives the rate of change of radius as required.

The last example included four variables, V, A, R and t. You used the chain rule to connect the rate of change. This is a connected rate of change problem (see Section 4.4).

Exercise 4E

1 In a study of the water loss of picked leaves the mass M grams of a single leaf was measured at times t days after the leaf was picked. It was found that the rate of loss of mass was proportional to the mass M of the leaf.
Write down a differential equation for the rate of change of mass of the leaf.

2 A curve C has equation $y = \mathrm{f}(x)$, $y > 0$. At any point P on the curve, the gradient of C is proportional to the product of the x and the y coordinates of P. The point A with coordinates (4, 2) is on C and the gradient of C at A is $\frac{1}{2}$.

Show that $\dfrac{dy}{dx} = \dfrac{xy}{16}$.

3 Liquid is pouring into a container at a constant rate of $30\,\text{cm}^3\,\text{s}^{-1}$. At time t seconds liquid is leaking from the container at a rate of $\frac{2}{15}V\,\text{cm}^3\,\text{s}^{-1}$, where $V\,\text{cm}^3$ is the volume of liquid in the container at that time.

Show that $-15\dfrac{dV}{dt} = 2V - 450$

4 An electrically charged body loses its charge Q coulombs at a rate, measured in coulombs per second, proportional to the charge Q.
Write down a differential equation in terms of Q and t where t is the time in seconds since the body started to lose its charge.

5 The ice on a pond has a thickness x mm at a time t hours after the start of freezing. The rate of increase of x is inversely proportional to the square of x.
Write down a differential equation in terms of x and t.

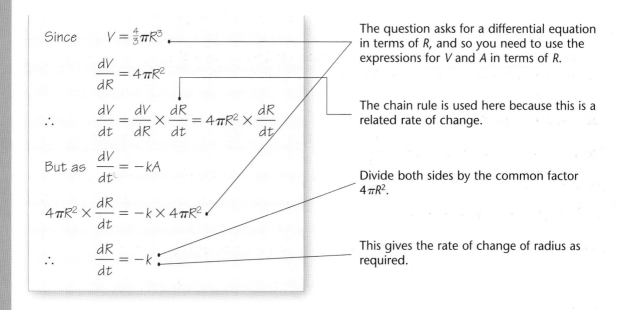

6 In another pond the amount of pondweed (P) grows at a rate proportional to the amount of pondweed already present in the pond. Pondweed is also removed by fish eating it at a constant rate of Q per unit of time.
Write down a differential equation relating P and t, where t is the time which has elapsed since the start of the observation.

7 A circular patch of oil on the surface of some water has radius r and the radius increases over time at a rate inversely proportional to the radius.
Write down a differential equation relating r and t, where t is the time which has elapsed since the start of the observation.

8 A metal bar is heated to a certain temperature, then allowed to cool down and it is noted that, at time t, the rate of loss of temperature is proportional to the difference in temperature between the metal bar, θ, and the temperature of its surroundings θ_0.
Write down a differential equation relating θ and t.

The next three questions involve connected rates of change.

9 Fluid flows out of a cylindrical tank with constant cross section. At time t minutes, $t > 0$, the volume of fluid remaining in the tank is $V\,\text{m}^3$. The rate at which the fluid flows in $\text{m}^3\,\text{min}^{-1}$ is proportional to the square root of V.
Show that the depth h metres of fluid in the tank satisfies the differential equation $\frac{dh}{dt} = -k\sqrt{h}$, where k is a positive constant.

10 At time t seconds the surface area of a cube is $A\,\text{cm}^2$ and the volume is $V\,\text{cm}^3$.
The surface area of the cube is expanding at a constant rate $2\,\text{cm}^2\,\text{s}^{-1}$.
Show that $\frac{dV}{dt} = \frac{1}{2}V^{\frac{1}{3}}$.

11 An inverted conical funnel is full of salt. The salt is allowed to leave by a small hole in the vertex. It leaves at a constant rate of $6\,\text{cm}^3\,\text{s}^{-1}$.
Given that the angle of the cone between the slanting edge and the vertical is 30 degrees, show that the volume of the salt is $\frac{1}{9}\pi h^3$, where h is the height of salt at time t seconds.
Show that the rate of change of the height of the salt in the funnel is inversely proportional to h^2. Write down the differential equation relating h and t.

Mixed exercise 4F

1 The curve C is given by the equations

$$x = 4t - 3,\ y = \frac{8}{t^2},\ t > 0$$

where t is a parameter.

At A, $t = 2$. The line l is the normal to C at A.

a Find $\frac{dy}{dx}$ in terms of t. **b** Hence find an equation of l. E

2 The curve C is given by the equations $x = 2t$, $y = t^2$, where t is a parameter. Find an equation of the normal to C at the point P on C where $t = 3$. **E**

3 The curve C has parametric equations

$$x = t^3,\ y = t^2,\ t > 0$$

Find an equation of the tangent to C at A (1, 1). **E**

4 A curve C is given by the equations

$$x = 2\cos t + \sin 2t,\ y = \cos t - 2\sin 2t,\ 0 < t < \pi$$

where t is a parameter.

a Find $\frac{dx}{dt}$ and $\frac{dy}{dt}$ in terms of t.

b Find the value of $\frac{dy}{dx}$ at the point P on C where $t = \frac{\pi}{4}$.

c Find an equation of the normal to the curve at P. **E**

5 A curve is given by $x = 2t + 3$, $y = t^3 - 4t$, where t is a parameter. The point A has parameter $t = -1$ and the line l is the tangent to C at A. The line l also cuts the curve at B.

a Show that an equation for l is $2y + x = 7$.

b Find the value of t at B. **E**

6 A Pancho car has value £V at time t years. A model for V assumes that the rate of decrease of V at time t is proportional to V. Form an appropriate differential equation for V. **E**

7 The curve shown has parametric equations

$$x = 5\cos\theta,\ y = 4\sin\theta,\ 0 \leqslant \theta < 2\pi$$

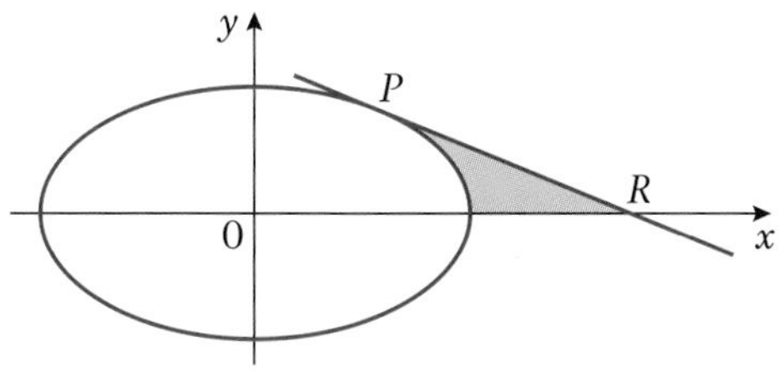

a Find the gradient of the curve at the point P at which $\theta = \frac{\pi}{4}$.

b Find an equation of the tangent to the curve at the point P.

c Find the coordinates of the point R where this tangent meets the x-axis. **E**

8 The curve C has parametric equations

$$x = 4\cos 2t,\ y = 3\sin t,\ -\frac{\pi}{2} < t < \frac{\pi}{2}$$

A is the point $(2, 1\frac{1}{2})$, and lies on C.

a Find the value of t at the point A.

b Find $\frac{dy}{dx}$ in terms of t.

c Show that an equation of the normal to C at A is $6y - 16x + 23 = 0$.

The normal at A cuts C again at the point B.

d Find the y-coordinate of the point B.

E

9 The diagram shows the curve C with parametric equations

$$x = a\sin^2 t,\ y = a\cos t,\ 0 \leqslant t \leqslant \frac{1}{2}\pi$$

where a is a positive constant. The point P lies on C and has coordinates $(\frac{3}{4}a, \frac{1}{2}a)$.

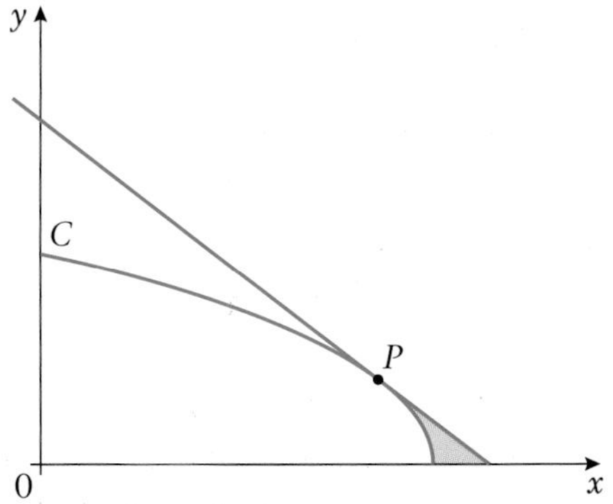

a Find $\frac{dy}{dx}$, giving your answer in terms of t.

b Find an equation of the tangent at P to C.

E

10 This graph shows part of the curve C with parametric equations

$$x = (t + 1)^2,\ y = \frac{1}{2}t^3 + 3,\ t \geqslant -1$$

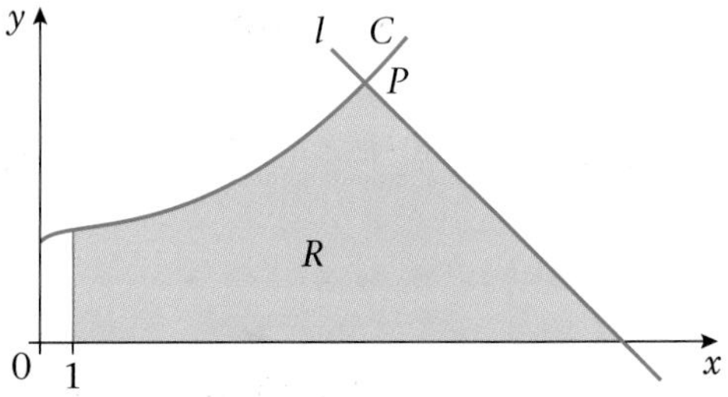

P is the point on the curve where $t = 2$. The line l is the normal to C at P.

Find the equation of l.

E

11 The diagram shows part of the curve C with parametric equations

$$x = t^2,\ y = \sin 2t,\ t \geqslant 0$$

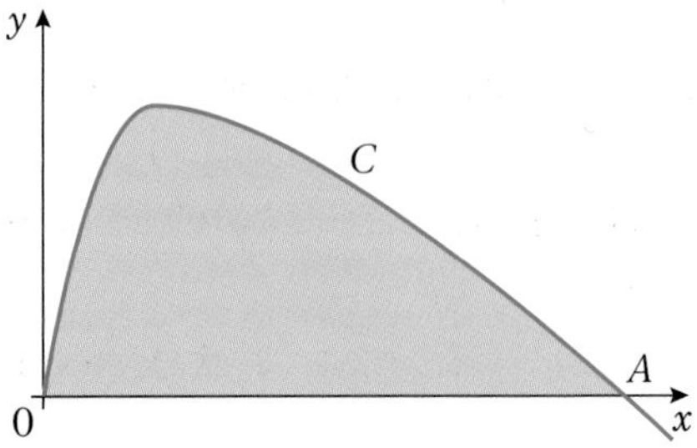

The point A is an intersection of C with the x-axis.

a Find, in terms of π, the x-coordinate of A.

b Find $\frac{dy}{dx}$ in terms of t, $t > 0$.

c Show that an equation of the tangent to C at A is $4x + 2\pi y = \pi^2$.

E

12 Find the gradient of the curve with equation

$$5x^2 + 5y^2 - 6xy = 13$$

at the point (1, 2). **E**

13 Given that $e^{2x} + e^{2y} = xy$, find $\frac{dy}{dx}$ in terms of x and y. **E**

14 Find the coordinates of the turning points on the curve $y^3 + 3xy^2 - x^3 = 3$. **E**

15 Given that $y(x + y) = 3$, evaluate $\frac{dy}{dx}$ when $y = 1$. **E**

16 **a** If $(1 + x)(2 + y) = x^2 + y^2$, find $\frac{dy}{dx}$ in terms of x and y.

b Find the gradient of the curve $(1 + x)(2 + y) = x^2 + y^2$ at each of the two points where the curve meets the y-axis.

c Show also that there are two points at which the tangents to this curve are parallel to the y-axis. **E**

17 A curve has equation $7x^2 + 48xy - 7y^2 + 75 = 0$. A and B are two distinct points on the curve and at each of these points the gradient of the curve is equal to $\frac{2}{11}$. Use implicit differentiation to show that $x + 2y = 0$ at the points A and B. **E**

18 Given that $y = x^x$, $x > 0$, $y > 0$, by taking logarithms show that

$$\frac{dy}{dx} = x^x(1 + \ln x)$$

E

19 **a** Given that $x = 2^t$, by using logarithms prove that

$$\frac{dx}{dt} = 2^t \ln 2$$

A curve C has parametric equations $x = 2^t$, $y = 3t^2$. The tangent to C at the point with coordinates (2, 3) cuts the x-axis at the point P.

b Find $\frac{dy}{dx}$ in terms of t.

c Calculate the x-coordinate of P, giving your answer to 3 decimal places. **E**

20 **a** Given that $a^x \equiv e^{kx}$, where a and k are constants, $a > 0$ and $x \in \mathbb{R}$, prove that $k = \ln a$.

b Hence, using the derivative of e^{kx}, prove that when $y = 2^x$

$$\frac{dy}{dx} = 2^x \ln 2.$$

c Hence deduce that the gradient of the curve with equation $y = 2^x$ at the point (2, 4) is ln 16. **E**

21 A population P is growing at the rate of 9% each year and at time t years may be approximated by the formula

$$P = P_0(1.09)^t,\ t \geqslant 0$$

where P is regarded as a continuous function of t and P_0 is the starting population at time $t = 0$.

a Find an expression for t in terms of P and P_0.

b Find the time T years when the population has doubled from its value at $t = 0$, giving your answer to 3 significant figures.

c Find, as a multiple of P_0, the rate of change of population $\frac{dP}{dt}$ at time $t = T$. **E**

Summary of key points

1 When a relation is described by parametric equations:

- You differentiate x and y with respect to the parameter t.
- Then you use the chain rule rearranged into the form $\frac{dy}{dx} = \frac{dy}{dt} \div \frac{dx}{dt}$.

2 When a relation is described by an implicit equation:

- Differentiate each term in turn, using the chain rule and product and quotient rules as appropriate.
- $\frac{d}{dx}(y^n) = ny^{n-1}\frac{dy}{dx}$ — By the chain rule.
- $\frac{d}{dx}(xy) = x\frac{d}{dx}(y) + y\frac{d}{dx}(x) = x\frac{dy}{dx} + y$ — By the product rule.

3 In an implicit equation:

- Note that when $f(y)$ is differentiated with respect to x it becomes $f'(y)\frac{dy}{dx}$.
- A product term such as $f(x).g(y)$ is differentiated by the product rule and becomes $f(x).g'(y)\frac{dy}{dx} + g(y).f'(x)$.

4 You can differentiate the function $f(x) = a^x$:

- If $y = a^x$, then $\frac{dy}{dx} = a^x \ln a$

5 You can use the chain rule once, or several times, to connect the rates of change in a question involving more than two variables.

6 You can set up simple differential equations from information given in context. This may involve using connected rates of change, or ideas of proportion.

5 Vectors

In this chapter you will learn how to use vectors to solve problems in two or three dimensions.

5.1 You need to know the difference between a scalar and a vector, and how to write down vectors and draw vector diagrams.

A scalar quantity can be described by using a single number (the *magnitude* or *size*).

■ **A vector quantity has both magnitude and direction.**

For example:

Scalar: The distance from P to Q is 100 metres.

Distance is a scalar.

Vector: From P to Q you go 100 metres north.

This is called the displacement from P to Q. Displacement is a vector.

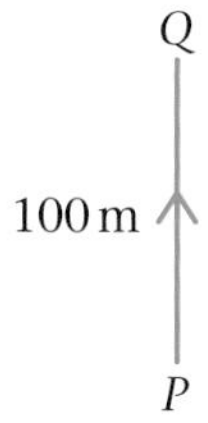

Scalar: A ship is sailing at 12 km h^{-1}.

Speed is a scalar.

Vector: A ship is sailing at 12 km h^{-1}, on a bearing of 060°.

This is called the velocity of the ship. Velocity is a vector.

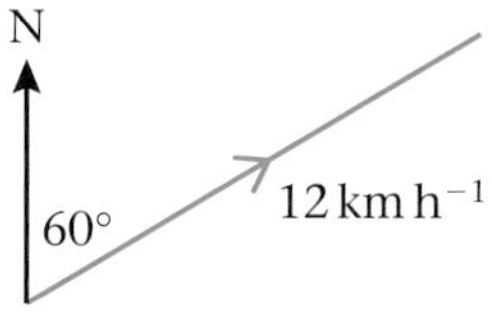

Example 1

Show on a diagram the displacement vector from P to Q, where Q is 500 m due north of P.

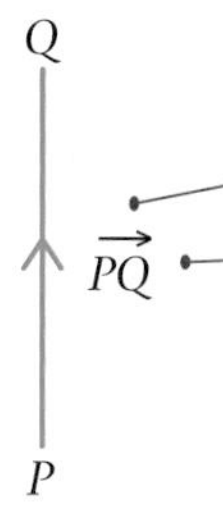

This is called a 'directed line segment'. The direction of the arrow shows the direction of the vector.

The vector is written as $\overrightarrow{PQ}$.

The length of the line segment PQ represents distance 500 m. In accurate diagrams a scale could be used (e.g. 1 cm represents 100 m).

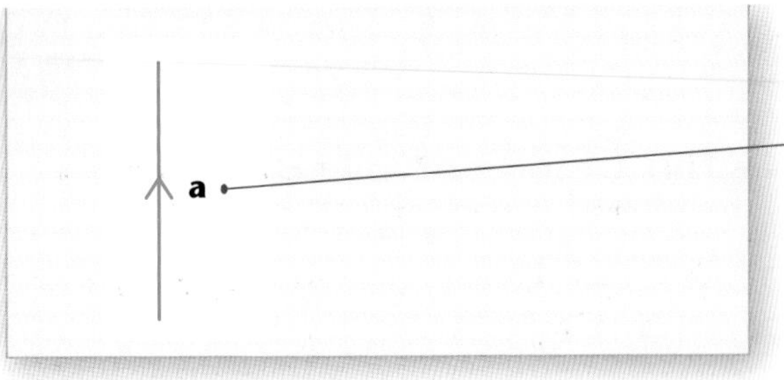

Sometimes, instead of using the endpoints P and Q, a small (lower case) letter is used.

In print, the small letter will be in **bold type**. In writing, you should underline the small letter to show it is a vector:

$\underline{a}$ or $\underset{\sim}{a}$

- **Vectors that are equal have both the same magnitude and the same direction.**

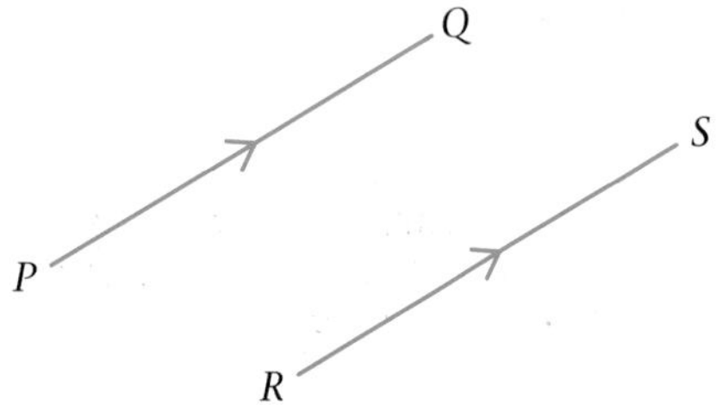

Here $\overrightarrow{PQ} = \overrightarrow{RS}$.

- **Two vectors are added using the 'triangle law'.**

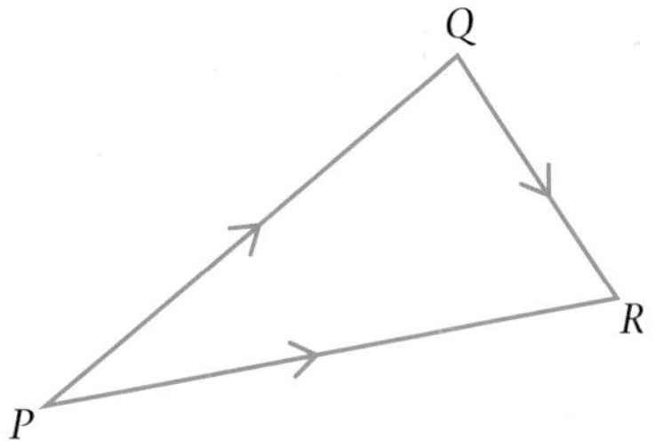

Hint: Think of displacement vectors.
If you travel from P to Q, then from Q to R, the resultant journey is P to R:

$$\overrightarrow{PQ} + \overrightarrow{QR} = \overrightarrow{PR}$$

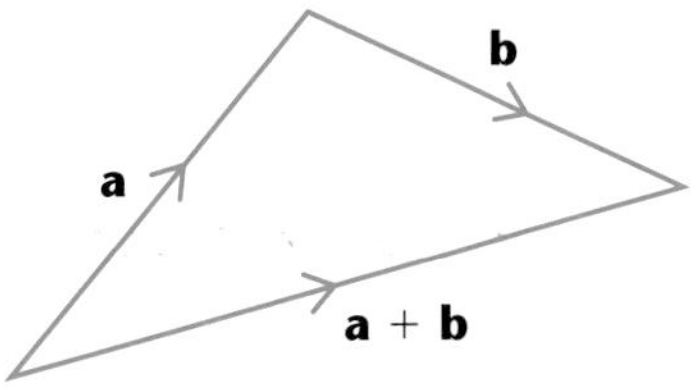

When you add the vectors **a** and **b**, the resultant vector **a** + **b** goes from 'the start of **a** to the finish of **b**'.

This is sometimes called the triangle law for vector addition.

Example 2

The diagram shows the vectors **a**, **b** and **c**. Draw another diagram to illustrate the vector addition **a** + **b** + **c**.

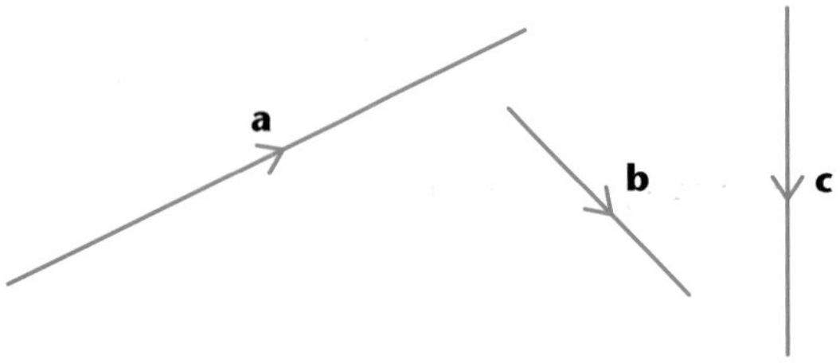

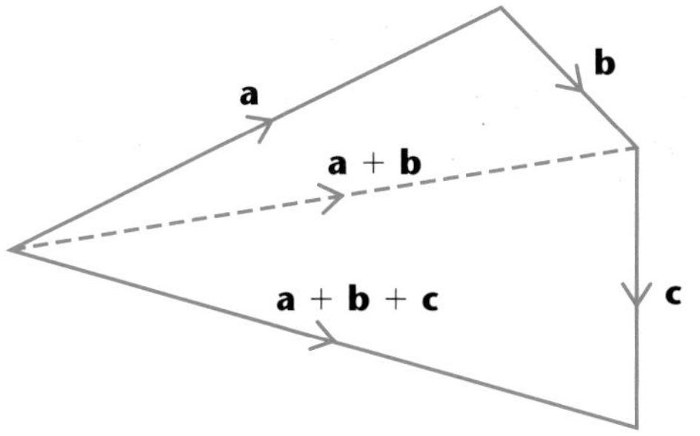

First use the triangle law for **a** + **b**, then use it again for (**a** + **b**) + **c**.

The resultant goes from the start of **a** to the finish of **c**.

■ **Adding the vectors $\overrightarrow{PQ}$ and $\overrightarrow{QP}$ gives the zero vector 0. $\overrightarrow{PQ} + \overrightarrow{QP} = \mathbf{0}$**

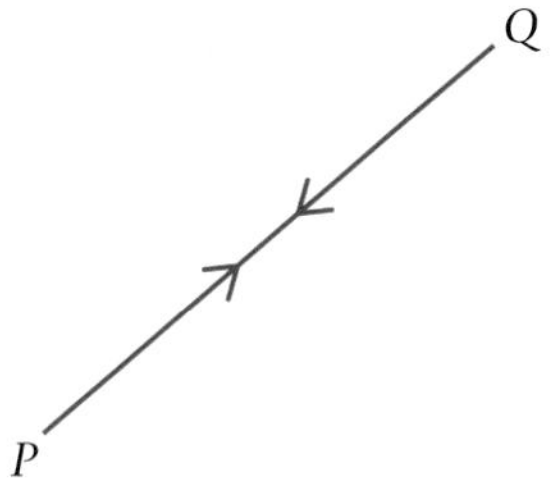

Hint: If you travel from P to Q, then back from Q to P, you are back where you started, so your displacement is zero.

The zero displacement vector is **0**.
It is printed in bold type, or underlined in written work.

You can also write $\overrightarrow{QP}$ as $-\overrightarrow{PQ}$.
So $\overrightarrow{PQ} + \overrightarrow{QP} = \mathbf{0}$ or $\overrightarrow{PQ} - \overrightarrow{PQ} = \mathbf{0}$.

■ **The modulus of a vector is another name for its magnitude.**

- **The modulus of the vector a is written as $|\mathbf{a}|$.**
- **The modulus of the vector $\overrightarrow{PQ}$ is written as $|\overrightarrow{PQ}|$.**

Example 3

The vector **a** is directed due east and $|\mathbf{a}| = 12$. The vector **b** is directed due south and $|\mathbf{b}| = 5$. Find $|\mathbf{a} + \mathbf{b}|$.

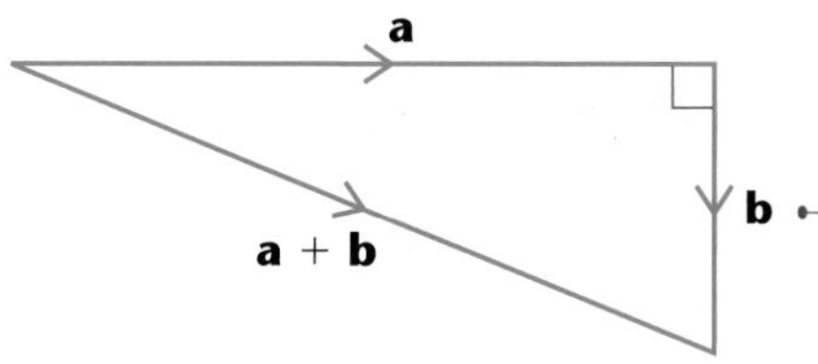

Use the triangle law for adding the vectors **a** and **b**.

$|\mathbf{a} + \mathbf{b}|^2 = 12^2 + 5^2 = 169$

Use Pythagoras' Theorem.

$|\mathbf{a} + \mathbf{b}| = 13$

Example 4

In the diagram, $\overrightarrow{QP} = \mathbf{a}$, $\overrightarrow{QR} = \mathbf{b}$, $\overrightarrow{QS} = \mathbf{c}$ and $\overrightarrow{RT} = \mathbf{d}$.

Find in terms of **a**, **b**, **c** and **d**:

a $\overrightarrow{PS}$ **b** $\overrightarrow{RP}$ **c** $\overrightarrow{PT}$ **d** $\overrightarrow{TS}$

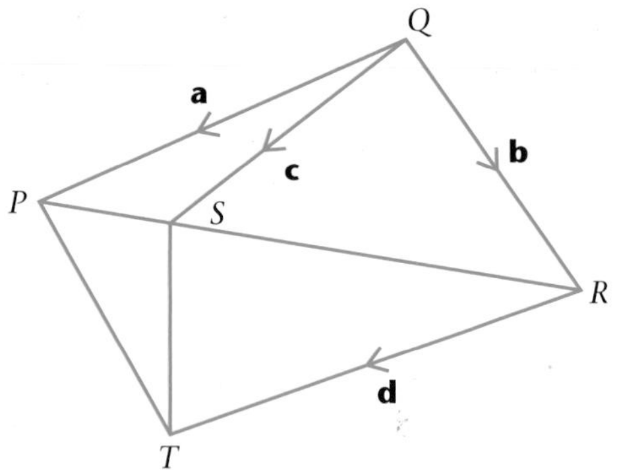

a $\overrightarrow{PS} = \overrightarrow{PQ} + \overrightarrow{QS} = -\mathbf{a} + \mathbf{c}$
$= \mathbf{c} - \mathbf{a}$

Add vectors using $\triangle PQS$.

b $\overrightarrow{RP} = \overrightarrow{RQ} + \overrightarrow{QP} = -\mathbf{b} + \mathbf{a}$
$= \mathbf{a} - \mathbf{b}$

Add vectors using $\triangle RQP$.

c $\overrightarrow{PT} = \overrightarrow{PR} + \overrightarrow{RT} = (\mathbf{b} - \mathbf{a}) + \mathbf{d}$
$= \mathbf{b} + \mathbf{d} - \mathbf{a}$

Add vectors using $\triangle PRT$.
Use $\overrightarrow{PR} = -\overrightarrow{RP} = -(\mathbf{a} - \mathbf{b}) = \mathbf{b} - \mathbf{a}$.

d $\overrightarrow{TS} = \overrightarrow{TR} + \overrightarrow{RS} = -\mathbf{d} + (\overrightarrow{RQ} + \overrightarrow{QS})$
$= \mathbf{d} + (-\mathbf{b} + \mathbf{c})$
$= \mathbf{c} - \mathbf{b} - \mathbf{d}$

Add vectors using $\triangle TRS$ and also $\triangle RQS$.

Exercise 5A

1 The diagram shows the vectors **a**, **b**, **c** and **d**. Draw a diagram to illustrate the vector addition **a** + **b** + **c** + **d**.

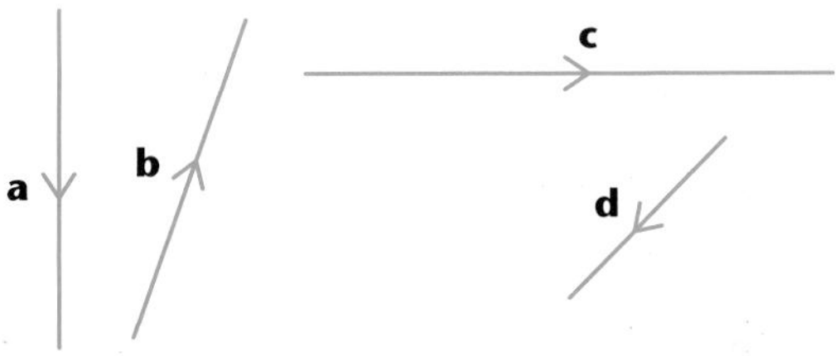

2 The vector **a** is directed due north and $|\mathbf{a}| = 24$. The vector **b** is directed due west and $|\mathbf{b}| = 7$. Find $|\mathbf{a} + \mathbf{b}|$.

3 The vector **a** is directed north-east and $|\mathbf{a}| = 20$. The vector **b** is directed south-east and $|\mathbf{b}| = 13$. Find $|\mathbf{a} + \mathbf{b}|$.

4 In the diagram, $\overrightarrow{PQ} = \mathbf{a}$, $\overrightarrow{QS} = \mathbf{b}$, $\overrightarrow{SR} = \mathbf{c}$ and $\overrightarrow{PT} = \mathbf{d}$. Find in terms of **a**, **b**, **c** and **d**:

a $\overrightarrow{QT}$

b $\overrightarrow{PR}$

c $\overrightarrow{TS}$

d $\overrightarrow{TR}$

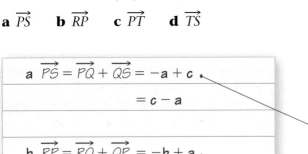

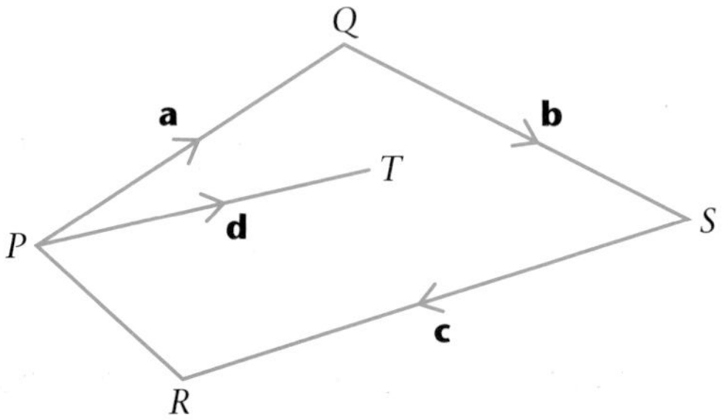

5 In the diagram, $\overrightarrow{WX} = \mathbf{a}$, $\overrightarrow{WY} = \mathbf{b}$ and $\overrightarrow{WZ} = \mathbf{c}$. It is given that $\overrightarrow{XY} = \overrightarrow{YZ}$.
Prove that $\mathbf{a} + \mathbf{c} = 2\mathbf{b}$.
($2\mathbf{b}$ is equivalent to $\mathbf{b} + \mathbf{b}$).

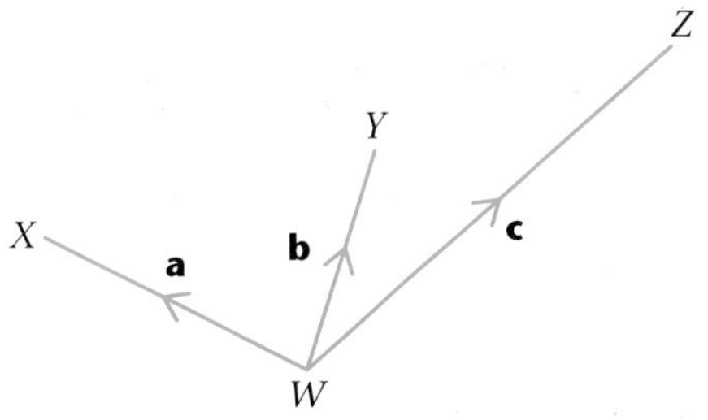

5.2 You need to be able to perform simple vector arithmetic, and to know the definition of a unit vector.

Example 5

The diagram shows the vector **a**. Draw diagrams to illustrate the vectors $3\mathbf{a}$ and $-2\mathbf{a}$.

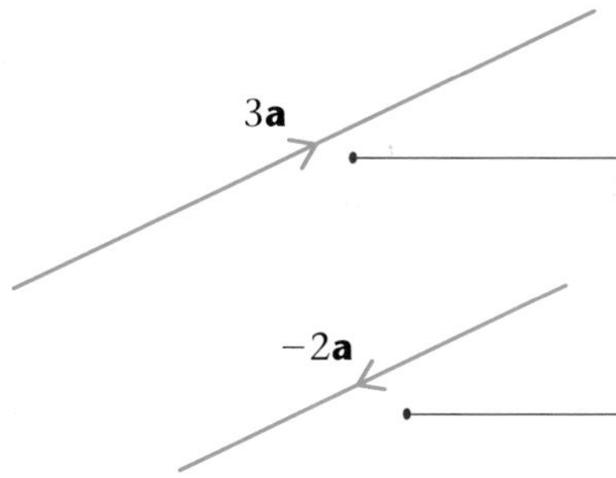

Vector $3\mathbf{a}$ is $\mathbf{a} + \mathbf{a} + \mathbf{a}$, so is in the same direction as **a** with 3 times its magnitude. The vector **a** has been multiplied by the scalar 3 (a scalar multiple).

Vector $-2\mathbf{a}$ is $-\mathbf{a} - \mathbf{a}$, so is in the opposite direction to **a** with 2 times its magnitude.

■ **Any vector parallel to the vector a may be written as λa, where λ is a non-zero scalar.**

Example 6

Show that the vectors $6\mathbf{a} + 8\mathbf{b}$ and $9\mathbf{a} + 12\mathbf{b}$ are parallel.

$9\mathbf{a} + 12\mathbf{b}$

$= \frac{3}{2}(6\mathbf{a} + 8\mathbf{b})$

∴ the vectors are parallel.

Here $\lambda = \dfrac{3}{2}$.

■ **Subtracting a vector is equivalent to 'adding a negative vector', so a – b is defined to be a + (–b).**

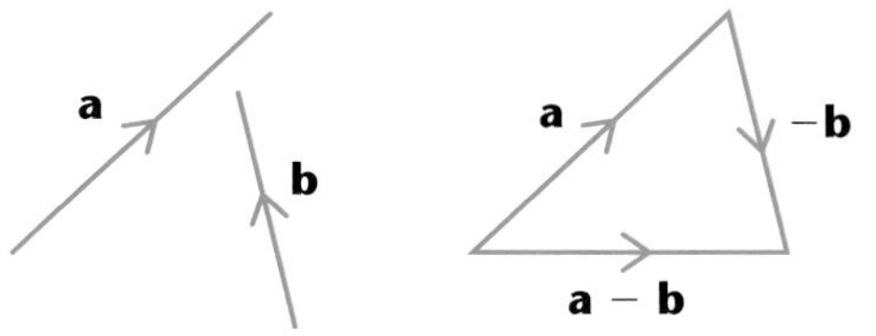

Hint: To subtract **b**, you reverse the direction of **b** then add.

■ **A unit vector is a vector which has magnitude (or modulus) 1 unit.**

Example 7

The vector **a** has magnitude 20 units. Write down a unit vector that is parallel to **a**.

The unit vector is $\frac{\mathbf{a}}{20}$ or $\frac{1}{20}\mathbf{a}$.

Divide **a** by the magnitude. In general, the unit vector is $\frac{\mathbf{a}}{|\mathbf{a}|}$.

■ **If $\lambda\mathbf{a} + \mu\mathbf{b} = \alpha\mathbf{a} + \beta\mathbf{b}$, and the non-zero vectors a and b are not parallel, then $\lambda = \alpha$ and $\mu = \beta$.**

The above result can be shown as follows:
$\lambda\mathbf{a} + \mu\mathbf{b} = \alpha\mathbf{a} + \beta\mathbf{b}$ can be written as $(\lambda - \alpha)\mathbf{a} = (\beta - \mu)\mathbf{b}$, but two vectors cannot be equal unless they are parallel or zero.
Since **a** and **b** are not parallel or zero, $(\lambda - \alpha) = 0$ and $(\beta - \mu) = 0$, so $\lambda = \alpha$ and $\beta = \mu$.

Example 8

Given that $5\mathbf{a} - 4\mathbf{b} = (2s + t)\mathbf{a} + (s - t)\mathbf{b}$, where **a** and **b** are non-zero, non-parallel vectors, find the values of the scalars s and t.

$2s + t = 5$
$s - t = -4$

Equate the **a** and **b** coefficients.

$3s = 1$

Solve simultaneously (add).

$s = \frac{1}{3}$

$t = 5 - 2s = 4\frac{1}{3}$

So $s = \frac{1}{3}$ and $t = 4\frac{1}{3}$.

Example 9

In the diagram, $\overrightarrow{PQ} = 3\mathbf{a}$, $\overrightarrow{QR} = \mathbf{b}$, $\overrightarrow{SR} = 4\mathbf{a}$ and $\overrightarrow{PX} = k\overrightarrow{PR}$.
Find, in terms of **a**, **b** and k:

a $\overrightarrow{PS}$ **b** $\overrightarrow{PX}$ **c** $\overrightarrow{SQ}$ **d** $\overrightarrow{SX}$

Use the fact that X lies on SQ to find the value of k.

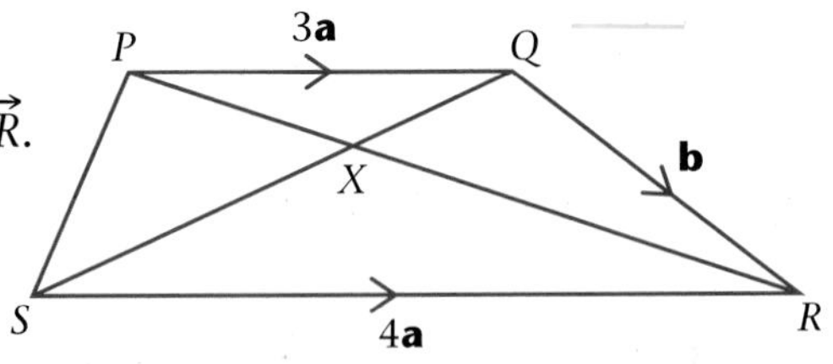

a $\overrightarrow{PS} = \overrightarrow{PR} + \overrightarrow{RS} = \overrightarrow{PQ} + \overrightarrow{QR} + \overrightarrow{RS}$

Add using the triangle law.

$= 3\mathbf{a} + \mathbf{b} - 4\mathbf{a} = \mathbf{b} - \mathbf{a}$

b $\overrightarrow{PR} = \overrightarrow{PQ} + \overrightarrow{QR} = 3\mathbf{a} + \mathbf{b}$

$\overrightarrow{PX} = k\overrightarrow{PR} = k(3\mathbf{a} + \mathbf{b})$

Find $\overrightarrow{PR}$ and use $\overrightarrow{PX} = k\overrightarrow{PR}$.

c $\overrightarrow{SQ} = 4\mathbf{a} - \mathbf{b}$ — Use the triangle law on $\triangle SRQ$.

d $\overrightarrow{SX} = \overrightarrow{SP} + \overrightarrow{PX}$

$= -(\mathbf{b} - \mathbf{a}) + k(3\mathbf{a} + \mathbf{b})$ — Use $\overrightarrow{SP} = -\overrightarrow{PS}$, and the answers to parts (a) and (b).

$= -\mathbf{b} + \mathbf{a} + k(3\mathbf{a} + \mathbf{b})$

$= (3k + 1)\mathbf{a} + (k - 1)\mathbf{b}$

X lies on SQ, so $\overrightarrow{SQ}$ and $\overrightarrow{SX}$ are parallel.

$(3k + 1)\mathbf{a} + (k - 1)\mathbf{b} = \lambda(4\mathbf{a} - \mathbf{b})$ — Use the fact that, for parallel vectors, one is a scalar multiple of the other.

$(3k + 1)\mathbf{a} + (k - 1)\mathbf{b} = 4\lambda\mathbf{a} - \lambda\mathbf{b}$

So $(3k + 1) = 4\lambda$ and $(k - 1) = -\lambda$ — **a** and **b** are non-parallel and non-zero, so equate coefficients.

$(3k + 1) = 4(1 - k)$

$k = \frac{3}{7}$ — Eliminate λ and solve for k.

Exercise 5B

1 In the triangle PQR, $\overrightarrow{PQ} = 2\mathbf{a}$ and $\overrightarrow{QR} = 2\mathbf{b}$. The mid-point of PR is M.
Find, in terms of **a** and **b**:

a $\overrightarrow{PR}$ **b** $\overrightarrow{PM}$ **c** $\overrightarrow{QM}$.

2 $ABCD$ is a trapezium with AB parallel to DC and $DC = 3AB$.
M is the mid-point of DC, $\overrightarrow{AB} = \mathbf{a}$ and $\overrightarrow{BC} = \mathbf{b}$.
Find, in terms of **a** and **b**:

a $\overrightarrow{AM}$ **b** $\overrightarrow{BD}$ **c** $\overrightarrow{MB}$ **d** $\overrightarrow{DA}$.

3 In each part, find whether the given vector is parallel to $\mathbf{a} - 3\mathbf{b}$:

a $2\mathbf{a} - 6\mathbf{b}$ **b** $4\mathbf{a} - 12\mathbf{b}$ **c** $\mathbf{a} + 3\mathbf{b}$

d $3\mathbf{b} - \mathbf{a}$ **e** $9\mathbf{b} - 3\mathbf{a}$ **f** $\frac{1}{2}\mathbf{a} - \frac{2}{3}\mathbf{b}$

4 The non-zero vectors **a** and **b** are not parallel. In each part, find the value of λ and the value of μ:

a $\mathbf{a} + 3\mathbf{b} = 2\lambda\mathbf{a} - \mu\mathbf{b}$

b $(\lambda + 2)\mathbf{a} + (\mu - 1)\mathbf{b} = \mathbf{0}$

c $4\lambda\mathbf{a} - 5\mathbf{b} - \mathbf{a} + \mu\mathbf{b} = \mathbf{0}$

d $(1 + \lambda)\mathbf{a} + 2\lambda\mathbf{b} = \mu\mathbf{a} + 4\mu\mathbf{b}$

e $(3\lambda + 5)\mathbf{a} + \mathbf{b} = 2\mu\mathbf{a} + (\lambda - 3)\mathbf{b}$

5 In the diagram, $\overrightarrow{OA} = \mathbf{a}$, $\overrightarrow{OB} = \mathbf{b}$ and C divides AB in the ratio 5:1.

a Write down, in terms of **a** and **b**, expressions for $\overrightarrow{AB}$, $\overrightarrow{AC}$ and $\overrightarrow{OC}$.

Given that $\overrightarrow{OE} = \lambda\mathbf{b}$, where λ is a scalar:

b Write down, in terms of **a**, **b** and λ, an expression for $\overrightarrow{CE}$.

Given that $\overrightarrow{OD} = \mu(\mathbf{b} - \mathbf{a})$, where μ is a scalar:

c Write down, in terms of **a**, **b**, λ and μ, an expression for $\overrightarrow{ED}$.

Given also that E is the mid-point of CD:

d Deduce the values of λ and μ.

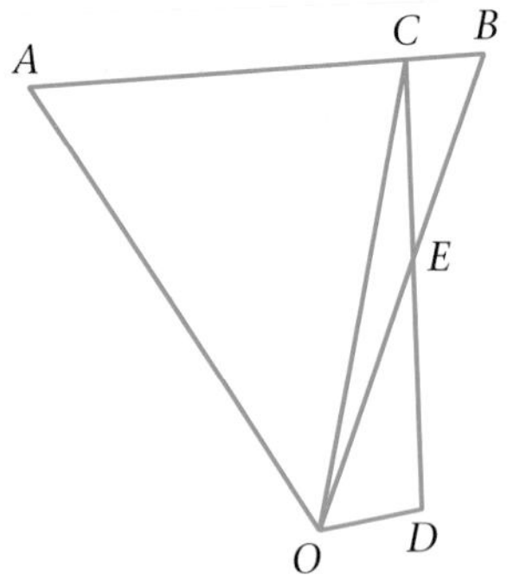

E

6 In the diagram $\overrightarrow{OA} = \mathbf{a}$, $\overrightarrow{OB} = \mathbf{b}$, $3\overrightarrow{OC} = 2\overrightarrow{OA}$ and $4\overrightarrow{OD} = 7\overrightarrow{OB}$.

The line DC meets the line AB at E.

a Write down, in terms of **a** and **b**, expressions for
i $\overrightarrow{AB}$ **ii** $\overrightarrow{DC}$

Given that $\overrightarrow{DE} = \lambda\overrightarrow{DC}$ and $\overrightarrow{EB} = \mu\overrightarrow{AB}$ where λ and μ are constants:

b Use $\triangle EBD$ to form an equation relating to **a**, **b**, λ and μ.

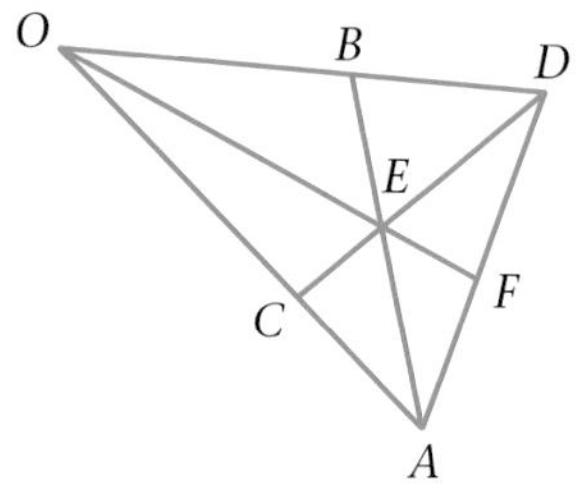

Hence:

c Show that $\lambda = \frac{9}{13}$. **d** Find the exact value of μ.

e Express $\overrightarrow{OE}$ in terms of **a** and **b**.

The line $\overrightarrow{OE}$ produced meets the line AD at F.
Given that $\overrightarrow{OF} = k\overrightarrow{OE}$ where k is a constant and that $\overrightarrow{AF} = \frac{1}{10}(7\mathbf{b} - 4\mathbf{a})$:

f Find the value of k.

E

7 In $\triangle OAB$, P is the mid-point of AB and Q is the point on OP such that $Q = \frac{3}{4}P$. Given that $\overrightarrow{OA} = \mathbf{a}$ and $\overrightarrow{OB} = \mathbf{b}$, find, in terms of **a** and **b**:

a $\overrightarrow{AB}$ **b** $\overrightarrow{OP}$ **c** $\overrightarrow{OQ}$ **d** $\overrightarrow{AQ}$

The point R on OB is such that $OR = kOB$, where $0 < k < 1$.

e Find, in terms of **a**, **b** and k, the vector $\overrightarrow{AR}$.

Given that AQR is a straight line:

f Find the ratio in which Q divides AR and the value of k.

E

8 In the figure $OE : EA = 1 : 2$, $AF : FB = 3 : 1$ and $OG : OB = 3 : 1$. The vector $\overrightarrow{OA} = \mathbf{a}$ and the vector $\overrightarrow{OB} = \mathbf{b}$.

Find, in terms of **a**, **b** or **a** and **b**, expressions for:

a $\overrightarrow{OE}$ **b** $\overrightarrow{OF}$ **c** $\overrightarrow{EF}$

d $\overrightarrow{BG}$ **e** $\overrightarrow{FB}$ **f** $\overrightarrow{FG}$

g Use your results in **c** and **f** to show that the points E, F and G are collinear and find the ratio $EF{:}FG$.

h Find $\overrightarrow{EB}$ and $\overrightarrow{AG}$ and hence prove that EB is parallel to AG.

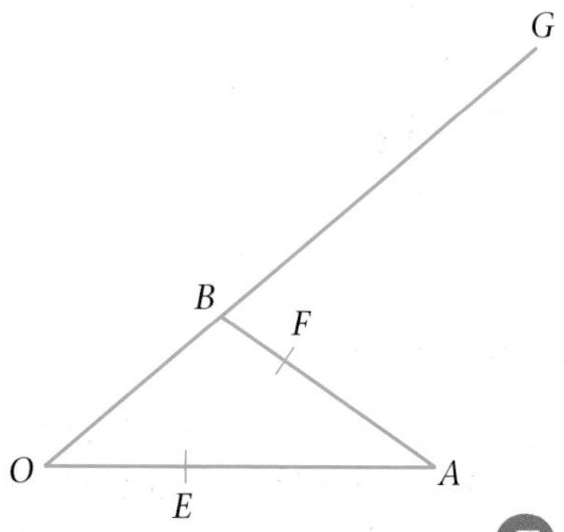

E

5.3 You need to be able to use vectors to describe the position of a point in two or three dimensions.

■ **The position vector of a point A is the vector $\overrightarrow{OA}$, where O is the origin. $\overrightarrow{OA}$ is usually written as vector $\mathbf{a}$.**

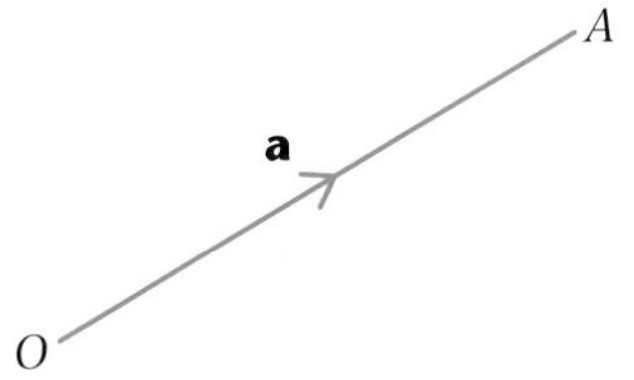

$\overrightarrow{OA} = \mathbf{a}$

■ **$\overrightarrow{AB} = \mathbf{b} - \mathbf{a}$, where $\mathbf{a}$ and $\mathbf{b}$ are the position vectors of A and B respectively.**

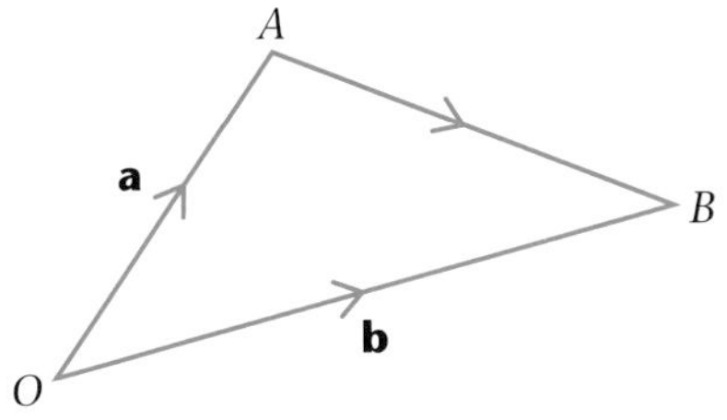

Hint: Use the triangle law to give
$\overrightarrow{AB} = \overrightarrow{AO} + \overrightarrow{OB} = -\mathbf{a} + \mathbf{b}$
So $\overrightarrow{AB} = \mathbf{b} - \mathbf{a}$

Example 10

In the diagram the points A and B have position vectors $\mathbf{a}$ and $\mathbf{b}$ respectively (referred to the origin O). The point P divides AB in the ratio $1:2$.

Find the position vector of P.

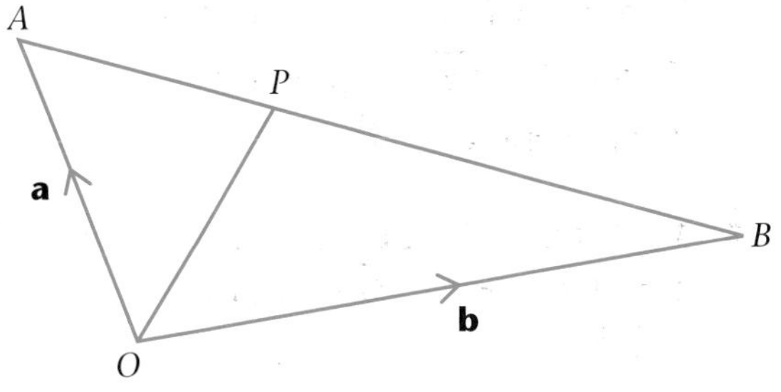

$\overrightarrow{AB} = \mathbf{b} - \mathbf{a}$

$\overrightarrow{OP} = \overrightarrow{OA} + \overrightarrow{AP}$ — $\overrightarrow{OP}$ is the position vector of P.

$\overrightarrow{AP} = \frac{1}{3}(\mathbf{b} - \mathbf{a})$ — Use the 1:2 ratio (AP is one third of AB).

$\overrightarrow{OP} = \mathbf{a} + \frac{1}{3}(\mathbf{b} - \mathbf{a})$

$\overrightarrow{OP} = \frac{2}{3}\mathbf{a} + \frac{1}{3}\mathbf{b}$ — You could write $\mathbf{p} = \frac{2}{3}\mathbf{a} + \frac{1}{3}\mathbf{b}$.

Exercise 5C

1 The points A and B have position vectors $\mathbf{a}$ and $\mathbf{b}$ respectively (referred to the origin O).
The point P divides AB in the ratio 1:5.
Find, in terms of $\mathbf{a}$ and $\mathbf{b}$, the position vector of P.

2 The points A, B and C have position vectors **a**, **b** and **c** respectively (referred to the origin O). The point P is the mid-point of AB.
Find, in terms of **a**, **b** and **c**, the vector $\overrightarrow{PC}$.

3 $OABCDE$ is a regular hexagon. The points A and B have position vectors **a** and **b** respectively, referred to the origin O.
Find, in terms of **a** and **b**, the position vectors of C, D and E.

5.4 You need to know how to write down and use the cartesian components of a vector in two dimensions.

■ **The vectors i and j are unit vectors parallel to the x-axis and the y-axis, and in the direction of x increasing and y increasing, respectively.**

Example 11

The points A and B in the diagram have coordinates (3, 4) and (11, 2) respectively.
Find, in terms of **i** and **j**:

a the position vector of A **b** the position vector of B **c** the vector $\overrightarrow{AB}$

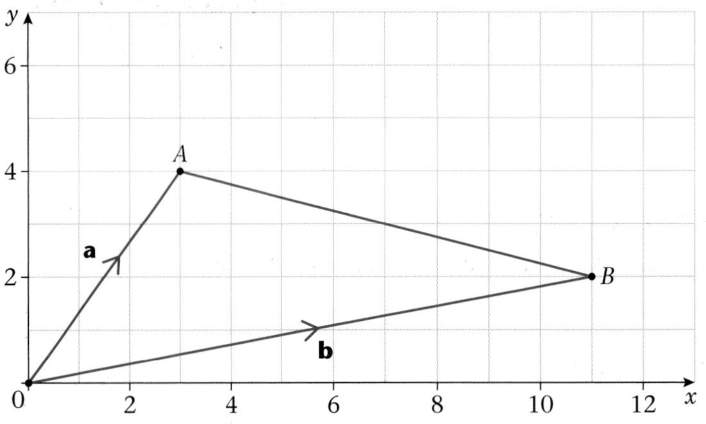

a $\mathbf{a} = \overrightarrow{OA} = 3\mathbf{i} + 4\mathbf{j}$

b $\mathbf{b} = \overrightarrow{OB} = 11\mathbf{i} + 2\mathbf{j}$

c $\overrightarrow{AB} = \mathbf{b} - \mathbf{a}$

$= (11\mathbf{i} + 2\mathbf{j}) - (3\mathbf{i} + 4\mathbf{j})$

$= 8\mathbf{i} - 2\mathbf{j}$

i goes 1 unit 'across', **j** goes 1 unit 'up':

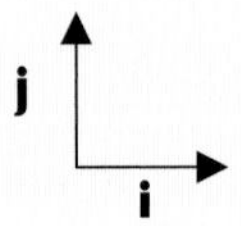

You can see from the diagram that the vector $\overrightarrow{AB}$ goes 8 units 'across' and 2 units 'down'.

■ **You can write a vector with cartesian components as a column matrix:**

$$x\mathbf{i} + y\mathbf{j} = \begin{pmatrix} x \\ y \end{pmatrix}$$

Hint: This standard notation is easy to read and also avoids the need to write out lengthy expressions with **i** and **j** terms.

Example 12

Given that $\mathbf{a} = 2\mathbf{i} + 5\mathbf{j}$, $\mathbf{b} = 12\mathbf{i} - 10\mathbf{j}$ and $\mathbf{c} = -3\mathbf{i} + 9\mathbf{j}$, find $\mathbf{a} + \mathbf{b} + \mathbf{c}$, using column matrix notation in your working.

$$\mathbf{a} + \mathbf{b} + \mathbf{c} = \begin{pmatrix} 2 \\ 5 \end{pmatrix} + \begin{pmatrix} 12 \\ -10 \end{pmatrix} + \begin{pmatrix} -3 \\ 9 \end{pmatrix}$$

$$= \begin{pmatrix} 11 \\ 4 \end{pmatrix}$$

Add the numbers in the top line to get 11 (the x component), and the bottom line to get 4 (the y component). This is $11\mathbf{i} + 4\mathbf{j}$.

■ **The modulus (or magnitude) of $x\mathbf{i} + y\mathbf{j}$ is $\sqrt{x^2 + y^2}$**

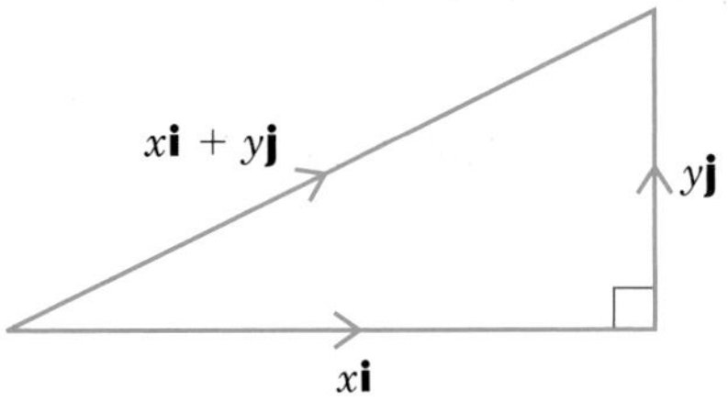

Hint: From Pythagoras' Theorem, the magnitude of $x\mathbf{i} + y\mathbf{j}$, represented by the hypotenuse, is $\sqrt{x^2 + y^2}$.

Example 13

The vector **a** is equal to $5\mathbf{i} - 12\mathbf{j}$.
Find $|\mathbf{a}|$, and find a unit vector in the same direction as **a**.

$$|\mathbf{a}| = \sqrt{5^2 + (-12)^2} = \sqrt{169} = 13$$

Unit vector is $\dfrac{\mathbf{a}}{|\mathbf{a}|}$

Look back to Section 5.2.

$$= \frac{5\mathbf{i} - 12\mathbf{j}}{13}$$

$$= \tfrac{1}{13}(5\mathbf{i} - 12\mathbf{j})$$

or $\tfrac{5}{13}\mathbf{i} - \tfrac{12}{13}\mathbf{j}$

or $\tfrac{1}{13}\begin{pmatrix} 5 \\ -12 \end{pmatrix}$

Example 14

Given that $\mathbf{a} = 5\mathbf{i} + \mathbf{j}$ and $\mathbf{b} = -2\mathbf{i} - 4\mathbf{j}$, find the exact value of $|2\mathbf{a} + \mathbf{b}|$.

$$2a + b = 2\begin{pmatrix}5\\1\end{pmatrix} + \begin{pmatrix}-2\\-4\end{pmatrix}$$

$$= \begin{pmatrix}10\\2\end{pmatrix} + \begin{pmatrix}-2\\-4\end{pmatrix}$$

$$= \begin{pmatrix}8\\-2\end{pmatrix}$$

$$|2a + b| = \sqrt{8^2 + (-2)^2}$$

$$= \sqrt{68}$$

$$= \sqrt{4}\sqrt{17}$$

$$= 2\sqrt{17}$$

You must give the answer as a surd because the question asks for an exact answer.

Exercise 5D

1 Given that $\mathbf{a} = 9\mathbf{i} + 7\mathbf{j}$, $\mathbf{b} = 11\mathbf{i} - 3\mathbf{j}$ and $\mathbf{c} = -8\mathbf{i} - \mathbf{j}$, find:

a $\mathbf{a} + \mathbf{b} + \mathbf{c}$

b $2\mathbf{a} - \mathbf{b} + \mathbf{c}$

c $2\mathbf{b} + 2\mathbf{c} - 3\mathbf{a}$

(Use column matrix notation in your working.)

2 The points A, B and C have coordinates $(3, -1)$, $(4, 5)$ and $(-2, 6)$ respectively, and O is the origin.

Find, in terms of $\mathbf{i}$ and $\mathbf{j}$:

a the position vectors of A, B and C

b $\overrightarrow{AB}$

c $\overrightarrow{AC}$

Find, in surd form:

d $|\overrightarrow{OC}|$

e $|\overrightarrow{AB}|$

f $|\overrightarrow{AC}|$

3 Given that $\mathbf{a} = 4\mathbf{i} + 3\mathbf{j}$, $\mathbf{b} = 5\mathbf{i} - 12\mathbf{j}$, $\mathbf{c} = -7\mathbf{i} + 24\mathbf{j}$ and $\mathbf{d} = \mathbf{i} - 3\mathbf{j}$, find a unit vector in the direction of **a**, **b**, **c** and **d**.

4 Given that $\mathbf{a} = 5\mathbf{i} + \mathbf{j}$ and $\mathbf{b} = \lambda\mathbf{i} + 3\mathbf{j}$, and that $|3\mathbf{a} + \mathbf{b}| = 10$, find the possible values of λ.

5.5 You need to know how to use cartesian coordinates in three dimensions.

Cartesian coordinate axes in three dimensions are usually called x, y and z axes, each being at right angles to each of the others.

The coordinates of a point in three dimensions are written as (x, y, z).

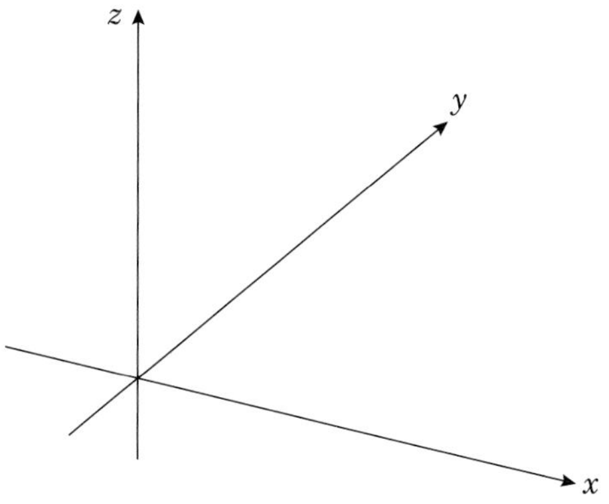

Hint: To visualise this, think of the x and y axes being drawn on a flat surface and the z axis sticking up from the surface.

Example 15

Find the distance between the points $P(4, 2, 5)$ and $Q(4, 2, -5)$.

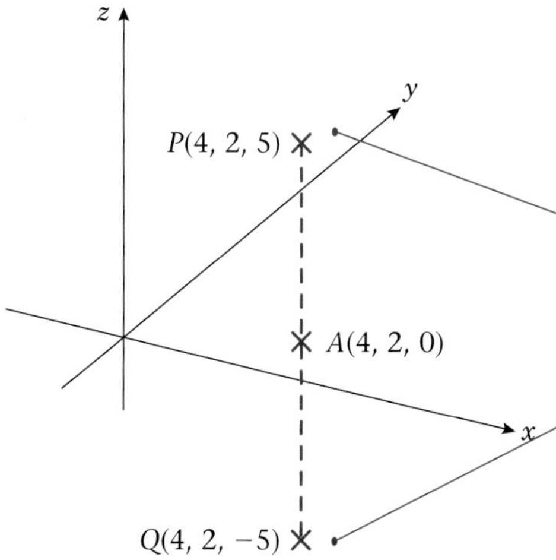

The distance between the points (4, 2, 5) and (4, 2, −5) is 5 + 5 = 10 units.

The point $A(4, 2, 0)$ is on the 'flat surface' (the xy plane).

(4, 2, 5) is 5 units 'above the surface'.

(4, 2, −5) is 5 units 'below the surface'.

So the line joining these 2 points is parallel to the z-axis.

Example 16

Find the distance from the origin to the point $P(4, 2, 5)$.

Let A be the point $(4, 2, 0)$.

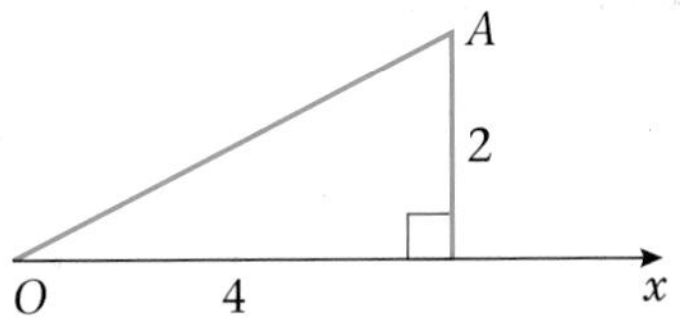

$OA^2 = 4^2 + 2^2$ — Use Pythagoras' Theorem in the xy plane.

$OA = \sqrt{(4^2 + 2^2)}$

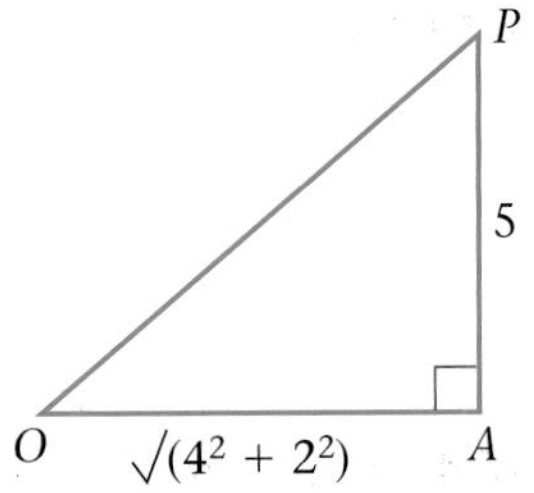

Next look at $\triangle OAP$, with OA on the xy plane and AP parallel to the z-axis.

Use Pythagoras' Theorem again.

$OP = \sqrt{(OA^2 + 5^2)}$

$OP = \sqrt{(4^2 + 2^2 + 5^2)}$ — Notice this method just gives the three-dimensional version of Pythagoras' Theorem.

$OP = \sqrt{45} = \sqrt{9}\ \sqrt{5} = 3\sqrt{5}.$

■ **The distance from the origin to the point (x, y, z) is $\sqrt{x^2 + y^2 + z^2}$.**

Pythagoras' Theorem in three dimensions.

Example 17

Find the distance from the origin to the point $P(4, -7, -1)$.

$OA = \sqrt{4^2 + (-7)^2 + (-1)^2}$ — Straight from the formula.

$OA = \sqrt{16 + 49 + 1} = \sqrt{66}$

$= 8.12$ (2 d.p.)

■ **The distance between the points (x_1, y_1, z_1) and (x_2, y_2, z_2) is $\sqrt{(x_1 - x_2)^2 + (y_1 - y_2)^2 + (z_1 - z_2)^2}$.**

This is the three-dimensional version of the formula $\sqrt{(x_1 - x_2)^2 + (y_1 - y_2)^2}$.

Example 18

Find the distance between the points $A(1, 3, 4)$ and $B(8, 6, -5)$, giving your answer to 1 decimal place.

$AB = \sqrt{(1-8)^2 + (3-6)^2 + (4-(-5))^2}$	Straight from the formula.
$= \sqrt{(-7)^2 + (-3)^2 + (9)^2}$	
$= \sqrt{139} = 11.8$ (1 d.p.)	

Example 19

The coordinates of A and B are $(5, 0, 3)$ and $(4, 2, k)$ respectively.
Given that the distance from A to B is 3 units, find the possible values of k.

$AB = \sqrt{(5-4)^2 + (0-2)^2 + (3-k)^2} = 3$	
$\sqrt{1 + 4 + (9 - 6k + k^2)} = 3$	
$1 + 4 + 9 - 6k + k^2 = 9$	Square both sides of the equation.
$k^2 - 6k + 5 = 0$	
$(k-5)(k-1) = 0$	Solve to find the two possible values of k.
$k = 1$ or $k = 5$	

Exercise 5E

1 Find the distance from the origin to the point $P(2, 8, -4)$.

2 Find the distance from the origin to the point $P(7, 7, 7)$.

3 Find the distance between A and B when they have the following coordinates:

a $A(3, 0, 5)$ and $B(1, -1, 8)$

b $A(8, 11, 8)$ and $B(-3, 1, 6)$

c $A(3, 5, -2)$ and $B(3, 10, 3)$

d $A(-1, -2, 5)$ and $B(4, -1, 3)$

4 The coordinates of A and B are $(7, -1, 2)$ and $(k, 0, 4)$ respectively.
Given that the distance from A to B is 3 units, find the possible values of k.

5 The coordinates of A and B are $(5, 3, -8)$ and $(1, k, -3)$ respectively.
Given that the distance from A to B is $3\sqrt{10}$ units, find the possible values of k.

5.6 You can extend the two-dimensional vector results to three dimensions, using k as the unit vector parallel to the z-axis, in the direction of z increasing.

Extending the results gives the following key points:

- **The vectors i, j and k are unit vectors parallel to the x-axis, the y-axis and the z-axis and in the direction of x increasing, x increasing and z increasing, respectively.**
- **The vector $x\mathbf{i} + y\mathbf{j} + z\mathbf{k}$, may be written as a column matrix $\begin{pmatrix} x \\ y \\ z \end{pmatrix}$.**
- **The modulus (or magnitude) of $x\mathbf{i} + y\mathbf{j} + z\mathbf{k}$ is $\sqrt{x^2 + y^2 + z^2}$.**

Example 20

The points A and B have position vectors $4\mathbf{i} + 2\mathbf{j} + 7\mathbf{k}$ and $3\mathbf{i} + 4\mathbf{j} - 1\mathbf{k}$ respectively, and O is the origin. Find $|\overrightarrow{AB}|$ and show that $\triangle OAB$ is isosceles.

$|\overrightarrow{OA}| = \mathbf{a} = \begin{pmatrix} 4 \\ 2 \\ 7 \end{pmatrix}, |\overrightarrow{OB}| = \mathbf{b} = \begin{pmatrix} 3 \\ 4 \\ -1 \end{pmatrix}$ — Write down the position vectors of A and B.

$\overrightarrow{AB} = \mathbf{b} - \mathbf{a} = \begin{pmatrix} 3 \\ 4 \\ -1 \end{pmatrix} - \begin{pmatrix} 4 \\ 2 \\ 7 \end{pmatrix} = \begin{pmatrix} -1 \\ 2 \\ -8 \end{pmatrix}$ — Use $\overrightarrow{AB} = \mathbf{b} - \mathbf{a}$.

$|\overrightarrow{AB}| = \sqrt{(-1)^2 + 2^2 + (-8)^2} = \sqrt{69}$ — Use the vector magnitude formula.

$|\overrightarrow{OA}| = \sqrt{4^2 + 2^2 + 7^2} = \sqrt{69}$

$|\overrightarrow{OB}| = \sqrt{3^2 + 4^2 + (-1)^2} = \sqrt{26}$ — Find the lengths of the other sides OA and OB of $\triangle OAB$.

So $\triangle OAB$ is isosceles, with $AB = OA$.

Example 21

The points A and B have coordinates $(t, 5, t - 1)$ and $(2t, t, 3)$ respectively.

a Find $|\overrightarrow{AB}|$.

b By differentiating $|\overrightarrow{AB}|^2$, find the value of t for which $|\overrightarrow{AB}|$ is a minimum.

c Find the minimum value of $|\overrightarrow{AB}|$.

a $\mathbf{a} = \begin{pmatrix} t \\ 5 \\ t-1 \end{pmatrix}$ and $\mathbf{b} = \begin{pmatrix} 2t \\ t \\ 3 \end{pmatrix}$ — Write down the position vectors of A and B.

$\overrightarrow{AB} = \begin{pmatrix} 2t \\ t \\ 3 \end{pmatrix} - \begin{pmatrix} t \\ 5 \\ t-1 \end{pmatrix} = \begin{pmatrix} t \\ t-5 \\ 4-t \end{pmatrix}$ — Use $\overrightarrow{AB} = \mathbf{b} - \mathbf{a}$.

$|\overrightarrow{AB}| = \sqrt{t^2 + (t-5)^2 + (4-t)^2}$ — Use the vector magnitude formula.

$= \sqrt{t^2 + t^2 - 10t + 25 + 16 - 8t + t^2}$

$= \sqrt{3t^2 - 18t + 41}$

b $|\overrightarrow{AB}|^2 = 3t^2 - 18t + 41$ — Call this p, and differentiate.

$$\frac{dp}{dt} = 6t - 18$$

For a minimum, $\frac{dp}{dt} = 0$

so $6t - 18 = 0$ — Use the fact that $\frac{dp}{dt} = 0$ at a minimum.

$t = 3$

$\frac{d^2p}{dt^2} = 6$, positive, $\therefore$ minimum. — Use the fact that if the second derivative is positive, the value is a minimum.

c $|\overrightarrow{AB}| = \sqrt{3t^2 - 18t + 41}$ — Substitute $t = 3$ back into $|\overrightarrow{AB}|$.

$= \sqrt{27 - 54 + 41}$

$= \sqrt{14}$

Exercise 5F

1 Find the modulus of:

a $3\mathbf{i} + 5\mathbf{j} + \mathbf{k}$ **b** $4\mathbf{i} - 2\mathbf{k}$ **c** $\mathbf{i} + \mathbf{j} - \mathbf{k}$

d $5\mathbf{i} - 9\mathbf{j} - 8\mathbf{k}$ **e** $\mathbf{i} + 5\mathbf{j} - 7\mathbf{k}$

2 Given that $\mathbf{a} = \begin{pmatrix} 5 \\ 0 \\ 2 \end{pmatrix}$, $\mathbf{b} = \begin{pmatrix} 2 \\ 1 \\ -3 \end{pmatrix}$ and $\mathbf{c} = \begin{pmatrix} 7 \\ -4 \\ 2 \end{pmatrix}$, find in column matrix form:

a $\mathbf{a} + \mathbf{b}$ **b** $\mathbf{b} - \mathbf{c}$ **c** $\mathbf{a} + \mathbf{b} + \mathbf{c}$

d $3\mathbf{a} - \mathbf{c}$ **e** $\mathbf{a} - 2\mathbf{b} + \mathbf{c}$ **f** $|\mathbf{a} - 2\mathbf{b} + \mathbf{c}|$

3 The position vector of the point A is $2\mathbf{i} - 7\mathbf{j} + 3\mathbf{k}$ and $\overrightarrow{AB} = 5\mathbf{i} + 4\mathbf{j} - \mathbf{k}$. Find the position of the point B.

4 Given that $\mathbf{a} = t\mathbf{i} + 2\mathbf{j} + 3\mathbf{k}$, and that $|\mathbf{a}| = 7$, find the possible values of t.

5 Given that $\mathbf{a} = 5t\mathbf{i} + 2t\mathbf{j} + t\mathbf{k}$, and that $|\mathbf{a}| = 3\sqrt{10}$, find the possible values of t.

6 The points A and B have position vectors $\begin{pmatrix} 2 \\ 9 \\ t \end{pmatrix}$ and $\begin{pmatrix} 2t \\ 5 \\ 3t \end{pmatrix}$ respectively.

a Find $\overrightarrow{AB}$.

b Find, in terms of t, $|\overrightarrow{AB}|$.

c Find the value of t that makes $|\overrightarrow{AB}|$ a minimum.

d Find the minimum value of $|\overrightarrow{AB}|$.

7 The points A and B have position vectors $\begin{pmatrix} 2t+1 \\ t+1 \\ 3 \end{pmatrix}$ and $\begin{pmatrix} t+1 \\ 5 \\ 2 \end{pmatrix}$ respectively.

a Find $\overrightarrow{AB}$.

b Find, in terms of t, $|\overrightarrow{AB}|$.

c Find the value of t that makes $|\overrightarrow{AB}|$ a minimum.

d Find the minimum value of $|\overrightarrow{AB}|$.

5.7 You need to know the definition of the scalar product of two vectors (in either two or three dimensions), and how it can be used to find the angle between two vectors.

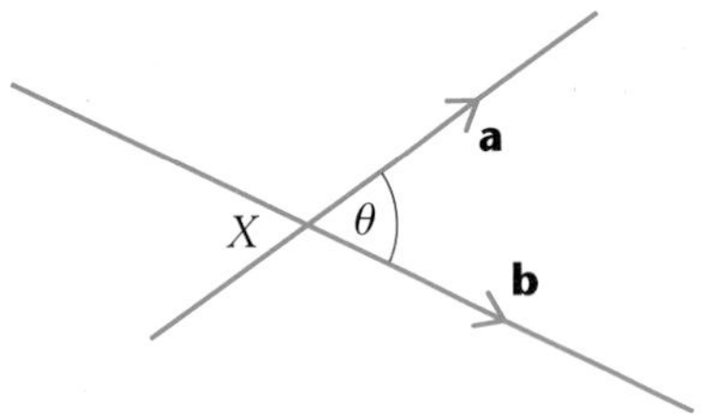

On the diagram, the angle between the vectors **a** and **b** is θ.

Notice that **a** and **b** are both directed **away from** the point X.

Example 22

Find the angle between the vectors **a** and **b** on the diagram:

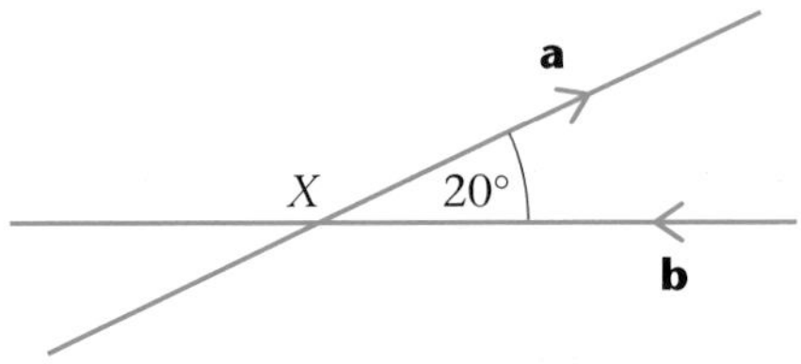

For the correct angle, **a** and **b** must both be pointing away from X, so re-draw to show this.

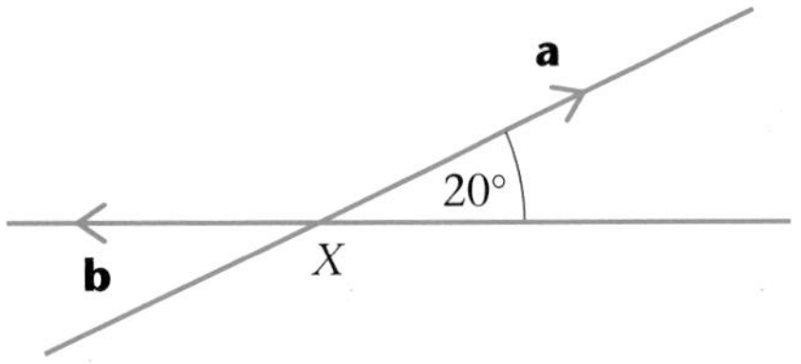

The angle between **a** and **b** is $180° - 20° = 160°$.

■ **The scalar product of two vectors a and b is written as a.b (say 'a dot b'), and defined by**

$$\mathbf{a}.\mathbf{b} = |\mathbf{a}||\mathbf{b}|\cos\theta$$

where θ is the angle between a and b.

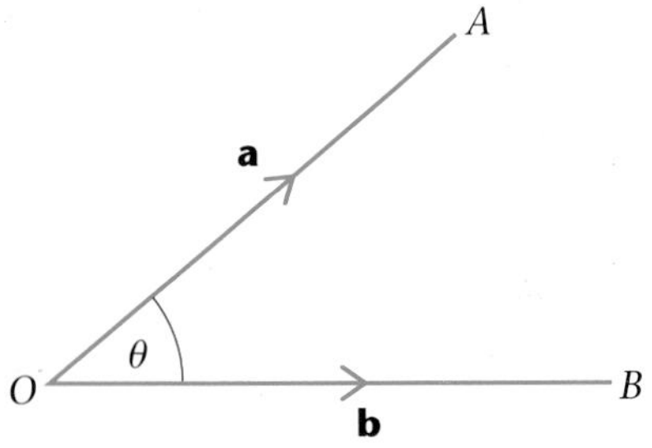

You can see from this diagram that if **a** and **b** are the position vectors of A and B, then the angle between **a** and **b** is $\angle AOB$.

■ **If a and b are the position vectors of the points A and B, then**

$$\cos AOB = \frac{\mathbf{a}.\mathbf{b}}{|\mathbf{a}||\mathbf{b}|}$$

If two vectors **a** and **b** are perpendicular, the angle between them is 90°.
Because $\cos 90° = 0$, then $\mathbf{a}.\mathbf{b} = |\mathbf{a}||\mathbf{b}|\cos 90° = 0$.

■ **The non-zero vectors a and b are perpendicular if and only if a.b = 0.**

Also, because $\cos 0° = 1$,

■ **If a and b are parallel, a.b = |a| |b|.**

- **In particular, a.a = |a|².**

$|\mathbf{a}|\,|\mathbf{b}|\cos 0°$

$|\mathbf{a}|\,|\mathbf{a}|\cos 0°$

Example 23

Find the value of

a **i.j** **b** **k.k** **c** $(4\mathbf{j}).\mathbf{k} + (3\mathbf{i}).(3\mathbf{i})$

a $\mathbf{i}.\mathbf{j} = 1 \times 1 \times \cos 90° = 0$

i and **j** are unit vectors (magnitude 1), and are perpendicular.

b $\mathbf{k}.\mathbf{k} = 1 \times 1 \times \cos 0° = 1$

k is a unit vector (magnitude 1), the angle between **k** and itself is 0°.

c $(4\mathbf{j}).\mathbf{k} + (3\mathbf{i}).(3\mathbf{i})$

$= (4 \times 1 \times \cos 90°) + (3 \times 3 \times \cos 0°)$

$= 0 + 9 = 9$

It can be shown that:

i $\mathbf{a}.\mathbf{b} = \mathbf{b}.\mathbf{a}$ and **ii** $\mathbf{a}.(\mathbf{b} + \mathbf{c}) = \mathbf{a}.\mathbf{b} + \mathbf{a}.\mathbf{c}$.

Because of these results, many processes which you are familiar with in ordinary algebra can be applied to the algebra of scalar products.

The proofs of results **i** and **ii** are shown below:

i $\mathbf{a}.\mathbf{b} = |\mathbf{a}|\,|\mathbf{b}|\cos\theta$, where θ is the angle between **a** and **b**.

$\mathbf{b}.\mathbf{a} = |\mathbf{b}|\,|\mathbf{a}|\cos\theta = |\mathbf{a}|\,|\mathbf{b}|\cos\theta$

So $\mathbf{a}.\mathbf{b} = \mathbf{b}.\mathbf{a}$.

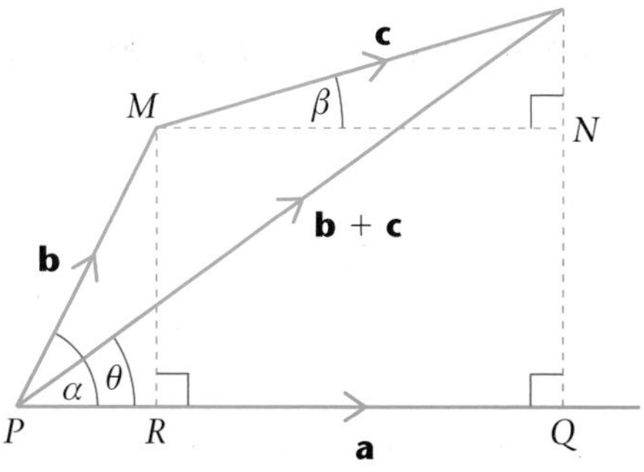

ii From the diagram,

$\mathbf{a}.(\mathbf{b} + \mathbf{c}) = |\mathbf{a}|\,|\mathbf{b} + \mathbf{c}|\cos\theta$

but $\cos\theta = \dfrac{PQ}{|\mathbf{b} + \mathbf{c}|}$, so $\mathbf{a}.(\mathbf{b} + \mathbf{c}) = |\mathbf{a}| \times PQ$

$\mathbf{a}.\mathbf{b} = |\mathbf{a}|\,|\mathbf{b}|\cos\alpha$, but $\cos\alpha = \dfrac{PR}{|\mathbf{b}|}$, so $\mathbf{a}.\mathbf{b} = |\mathbf{a}| \times PR$

$\mathbf{a}.\mathbf{c} = |\mathbf{a}|\,|\mathbf{c}|\cos\beta$,

Since MN is parallel to RQ, the angle between **c** and **a** is the same as that between **c** and MN, i.e. β.

but $\cos\beta = \dfrac{MN}{|\mathbf{c}|} = \dfrac{RQ}{|\mathbf{c}|}$,

so $\mathbf{a}.\mathbf{c} = |\mathbf{a}| \times RQ$

so $\mathbf{a}.(\mathbf{b} + \mathbf{c}) = |\mathbf{a}| \times PQ = |\mathbf{a}| \times (PR + RQ) = (|\mathbf{a}| \times PR) + (|\mathbf{a}| \times RQ) = \mathbf{a}.\mathbf{b} + \mathbf{a}.\mathbf{c}$

$\therefore \mathbf{a}.(\mathbf{b} + \mathbf{c}) = \mathbf{a}.\mathbf{b} + \mathbf{a}.\mathbf{c}$

Example 24

Given that $\mathbf{a} = \begin{pmatrix} a_1 \\ a_2 \\ a_3 \end{pmatrix}$ and $\mathbf{b} = \begin{pmatrix} b_1 \\ b_2 \\ b_3 \end{pmatrix}$ find $\mathbf{a}.\mathbf{b}$

$$\mathbf{a}.\mathbf{b} = (a_1\mathbf{i} + a_2\mathbf{j} + a_3\mathbf{k}).(b_1\mathbf{i} + b_2\mathbf{j} + b_3\mathbf{k})$$
$$= a_1\mathbf{i}.(b_1\mathbf{i} + b_2\mathbf{j} + b_3\mathbf{k})$$
$$+ a_2\mathbf{j}.(b_1\mathbf{i} + b_2\mathbf{j} + b_3\mathbf{k})$$
$$+ a_3\mathbf{k}.(b_1\mathbf{i} + b_2\mathbf{j} + b_3\mathbf{k})$$
$$= (a_1\mathbf{i}).(b_1\mathbf{i}) + (a_1\mathbf{i}).(b_2\mathbf{j}) + (a_1\mathbf{i}).(b_3\mathbf{k})$$
$$+ (a_2\mathbf{j}).(b_1\mathbf{i}) + (a_2\mathbf{j}).(b_2\mathbf{j}) + (a_2\mathbf{j}).(b_3\mathbf{k})$$
$$+ (a_3\mathbf{k}).(b_1\mathbf{i}) + (a_3\mathbf{k}).(b_2\mathbf{j})$$
$$+ (a_3\mathbf{k}).(b_3\mathbf{k})$$
$$= (a_1b_1)\mathbf{i}.\mathbf{i} + (a_1b_2)\mathbf{i}.\mathbf{j} + (a_1b_3)\mathbf{i}.\mathbf{k}$$
$$+ (a_2b_1)\mathbf{j}.\mathbf{i} + (a_2b_2)\mathbf{j}.\mathbf{j} + (a_2b_3)\mathbf{j}.\mathbf{k}$$
$$+ (a_3b_1)\mathbf{k}.\mathbf{i} + (a_3b_2)\mathbf{k}.\mathbf{j} + (a_3b_3)\mathbf{k}.\mathbf{k}$$
$$= a_1b_1 + a_2b_2 + a_3b_3$$

Use the results for parallel and perpendicular unit vectors:

$\mathbf{i}.\mathbf{i} = \mathbf{j}.\mathbf{j} = \mathbf{k}.\mathbf{k} = 1$

$\mathbf{i}.\mathbf{j} = \mathbf{i}.\mathbf{k} = \mathbf{j}.\mathbf{i} = \mathbf{j}.\mathbf{k} = \mathbf{k}.\mathbf{i} = \mathbf{k}.\mathbf{j} = 0$

The above example leads to a very simple formula for finding the scalar product of 2 vectors that are given in cartesian component form:

■ **If $\mathbf{a} = a_1\mathbf{i} + a_2\mathbf{j} + a_3\mathbf{k}$ and $\mathbf{b} = b_1\mathbf{i} + b_2\mathbf{j} + b_3\mathbf{k}$,**

$$\mathbf{a}.\mathbf{b} = \begin{pmatrix} a_1 \\ a_2 \\ a_3 \end{pmatrix}.\begin{pmatrix} b_1 \\ b_2 \\ b_3 \end{pmatrix} = a_1b_1 + a_2b_2 + a_3b_3$$

Example 25

Given that $\mathbf{a} = 8\mathbf{i} - 5\mathbf{j} - 4\mathbf{k}$ and $\mathbf{b} = 5\mathbf{i} + 4\mathbf{j} - \mathbf{k}$:

a Find $\mathbf{a}.\mathbf{b}$

b Find the angle between **a** and **b**, giving your answer in degrees to 1 decimal place.

a $\mathbf{a}.\mathbf{b} = \begin{pmatrix} 8 \\ -5 \\ -4 \end{pmatrix}.\begin{pmatrix} 5 \\ 4 \\ -1 \end{pmatrix}$ — Write in column matrix form.

$= (8 \times 5) + (-5 \times 4) + (-4 \times -1)$ — Use $\mathbf{a}.\mathbf{b} = a_1b_1 + a_2b_2 + a_3b_3$.

$= 40 - 20 + 4$

$= 24$

b $\mathbf{a}.\mathbf{b} = |\mathbf{a}|\,|\mathbf{b}|\cos\theta$ — Use the scalar product definition.

$|\mathbf{a}| = \sqrt{8^2 + (-5)^2 + (-4)^2} = \sqrt{105}$

$|\mathbf{b}| = \sqrt{5^2 + 4^2 + (-1)^2} = \sqrt{42}$ — Find the modulus of **a** and of **b**.

$\sqrt{105}\,\sqrt{42}\cos\theta = 24$ — Use $\mathbf{a}.\mathbf{b} = |\mathbf{a}|\,|\mathbf{b}|\cos\theta$.

$$\cos\theta = \frac{24}{\sqrt{105}\,\sqrt{42}}$$

$\theta = 68.8°$ (1 d.p.)

Example 26

Given that $\mathbf{a} = -\mathbf{i} + \mathbf{j} + 3\mathbf{k}$ and $\mathbf{b} = 7\mathbf{i} - 2\mathbf{j} + 2\mathbf{k}$, find the angle between **a** and **b**, giving your answer in degrees to 1 decimal place.

$$\mathbf{a}.\mathbf{b} = \begin{pmatrix}-1\\1\\3\end{pmatrix}.\begin{pmatrix}7\\-2\\2\end{pmatrix} = -7 - 2 + 6 = -3$$

$|\mathbf{a}| = \sqrt{(-1)^2 + 1^2 + 3^2} = \sqrt{11}$

$|\mathbf{b}| = \sqrt{(-7)^2 + (-2)^2 + 2^2} = \sqrt{57}$

For the scalar product formula, you need to find **a.b**, |**a**| and |**b**|.

$\sqrt{11}\,\sqrt{57}\cos\theta = -3$ — Use $\mathbf{a}.\mathbf{b} = |\mathbf{a}|\,|\mathbf{b}|\cos\theta$.

$$\cos\theta = \frac{-3}{\sqrt{11}\,\sqrt{57}}$$

The cosine is negative, so the angle is obtuse.

$\theta = 96.9°$ (1 d.p.)

Example 27

Given that the vectors $\mathbf{a} = 2\mathbf{i} - 6\mathbf{j} + \mathbf{k}$ and $\mathbf{b} = 5\mathbf{i} + 2\mathbf{j} + \lambda\mathbf{k}$ are perpendicular, find the value of λ.

$$\mathbf{a}.\mathbf{b} = \begin{pmatrix}2\\-6\\1\end{pmatrix}.\begin{pmatrix}5\\2\\\lambda\end{pmatrix}$$

$= 10 - 12 + \lambda$

$= -2 + \lambda$

Find the scalar product.

$-2 + \lambda = 0$ — For perpendicular vectors, the scalar product is zero.

$\lambda = 2$

Example 28

Given that $\mathbf{a} = -2\mathbf{i} + 5\mathbf{j} - 4\mathbf{k}$ and $\mathbf{b} = 4\mathbf{i} - 8\mathbf{j} + 5\mathbf{k}$, find a vector which is perpendicular to both **a** and **b**.

Let the required vector be $x\mathrm{i} + y\mathrm{j} + z\mathrm{k}$.

$$\mathbf{a}.\begin{pmatrix} x \\ y \\ z \end{pmatrix} = 0 \text{ and } \mathbf{b}.\begin{pmatrix} x \\ y \\ z \end{pmatrix} = 0$$

Both scalar products are zero.

$$\begin{pmatrix} -2 \\ 5 \\ -4 \end{pmatrix}.\begin{pmatrix} x \\ y \\ z \end{pmatrix} = 0 \text{ and } \begin{pmatrix} 4 \\ -8 \\ 5 \end{pmatrix}.\begin{pmatrix} x \\ y \\ z \end{pmatrix} = 0$$

$$-2x + 5y - 4z = 0$$

$$4x - 8y + 5z = 0$$

Let $z = 1$

Choose a (non-zero) value for z (or for x, or for y).

$$-2x + 5y = 4 \quad (\times 2)$$

$$4x - 8y = -5$$

$$-4x + 10y = 8$$

$$4x - 8y = -5$$

Adding, $2y = 3$

$$y = \tfrac{3}{2}$$

$$-2x + \tfrac{15}{2} = 4,\ 2x = \tfrac{7}{2}$$

$$x = \tfrac{7}{4}$$

So $x = \frac{7}{4}$, $y = \frac{3}{2}$ and $z = 1$

Solve simultaneously, by multiplying the first equation by 2, and eliminating x.

A possible vector is $\frac{7}{4}\mathrm{i} + \frac{3}{2}\mathrm{j} + \mathrm{k}$

Another possible vector is $4(\frac{7}{4}\mathrm{i} + \frac{3}{2}\mathrm{j} + \mathrm{k})$

$$= 7\mathrm{i} + 6\mathrm{j} + 4\mathrm{k}$$

You can multiply by a scalar constant to find another vector which is also perpendicular to both **a** and **b**.

Exercise 5G

1 The vectors $\mathbf{a}$ and $\mathbf{b}$ each have magnitude 3 units, and the angle between $\mathbf{a}$ and $\mathbf{b}$ is 60°. Find $\mathbf{a}.\mathbf{b}$.

2 In each part, find $\mathbf{a}.\mathbf{b}$:

a $\mathbf{a} = 5\mathbf{i} + 2\mathbf{j} + 3\mathbf{k}$, $\mathbf{b} = 2\mathbf{i} - \mathbf{j} - 2\mathbf{k}$

b $\mathbf{a} = 10\mathbf{i} - 7\mathbf{j} + 4\mathbf{k}$, $\mathbf{b} = 3\mathbf{i} - 5\mathbf{j} - 12\mathbf{k}$

c $\mathbf{a} = \mathbf{i} + \mathbf{j} - \mathbf{k}$, $\mathbf{b} = -\mathbf{i} - \mathbf{j} + 4\mathbf{k}$

d $\mathbf{a} = 2\mathbf{i} - \mathbf{k}$, $\mathbf{b} = 6\mathbf{i} - 5\mathbf{j} - 8\mathbf{k}$

e $\mathbf{a} = 3\mathbf{j} + 9\mathbf{k}$, $\mathbf{b} = \mathbf{i} + 12\mathbf{j} - 4\mathbf{k}$

3 In each part, find the angle between $\mathbf{a}$ and $\mathbf{b}$, giving your answer in degrees to 1 decimal place:

a $\mathbf{a} = 3\mathbf{i} + 7\mathbf{j}$, $\mathbf{b} = 5\mathbf{i} + \mathbf{j}$

b $\mathbf{a} = 2\mathbf{i} - 5\mathbf{j}$, $\mathbf{b} = 6\mathbf{i} + 3\mathbf{j}$

c $\mathbf{a} = \mathbf{i} - 7\mathbf{j} + 8\mathbf{k}$, $\mathbf{b} = 12\mathbf{i} + 2\mathbf{j} + \mathbf{k}$

d $\mathbf{a} = -\mathbf{i} - \mathbf{j} + 5\mathbf{k}$, $\mathbf{b} = 11\mathbf{i} - 3\mathbf{j} + 4\mathbf{k}$

e $\mathbf{a} = 6\mathbf{i} - 7\mathbf{j} + 12\mathbf{k}$, $\mathbf{b} = -2\mathbf{i} + \mathbf{j} + \mathbf{k}$

f $\mathbf{a} = 4\mathbf{i} + 5\mathbf{k}$, $\mathbf{b} = 6\mathbf{i} - 2\mathbf{j}$

g $\mathbf{a} = -5\mathbf{i} + 2\mathbf{j} - 3\mathbf{k}$, $\mathbf{b} = 2\mathbf{i} - 2\mathbf{j} + 11\mathbf{k}$

h $\mathbf{a} = \mathbf{i} + \mathbf{j} + \mathbf{k}$, $\mathbf{b} = \mathbf{i} - \mathbf{j} + \mathbf{k}$

4 Find the value, or values, of λ for which the given vectors are perpendicular:

a $3\mathbf{i} + 5\mathbf{j}$ and $\lambda\mathbf{i} + 6\mathbf{j}$

b $2\mathbf{i} + 6\mathbf{j} - \mathbf{k}$ and $\lambda\mathbf{i} - 4\mathbf{j} - 14\mathbf{k}$

c $3\mathbf{i} + \lambda\mathbf{j} - 8\mathbf{k}$ and $7\mathbf{i} - 5\mathbf{j} + \mathbf{k}$

d $9\mathbf{i} - 3\mathbf{j} + 5\mathbf{k}$ and $\lambda\mathbf{i} + \lambda\mathbf{j} + 3\mathbf{k}$

e $\lambda\mathbf{i} + 3\mathbf{j} - 2\mathbf{k}$ and $\lambda\mathbf{i} + \lambda\mathbf{j} + 5\mathbf{k}$

5 Find, to the nearest tenth of a degree, the angle that the vector $9\mathbf{i} - 5\mathbf{j} + 3\mathbf{k}$ makes with:

a the positive x-axis

b the positive y-axis

6 Find, to the nearest tenth of a degree, the angle that the vector $\mathbf{i} + 11\mathbf{j} - 4\mathbf{k}$ makes with:

a the positive y-axis

b the positive z-axis

7 The angle between the vectors $\mathbf{i} + \mathbf{j} + \mathbf{k}$ and $2\mathbf{i} + \mathbf{j} + \mathbf{k}$ is θ.
Calculate the exact value of $\cos\theta$.

8 The angle between the vectors $\mathbf{i} + 3\mathbf{j}$ and $\mathbf{j} + \lambda\mathbf{k}$ is 60°.

Show that $\lambda = \pm\sqrt{\frac{13}{5}}$.

9 Simplify as far as possible:

a $\mathbf{a}.(\mathbf{b} + \mathbf{c}) + \mathbf{b}.(\mathbf{a} - \mathbf{c})$, given that $\mathbf{b}$ is perpendicular to $\mathbf{c}$.

b $(\mathbf{a} + \mathbf{b}).(\mathbf{a} + \mathbf{b})$, given that $|\mathbf{a}| = 2$ and $|\mathbf{b}| = 3$.

c $(\mathbf{a} + \mathbf{b}).(2\mathbf{a} - \mathbf{b})$, given that $\mathbf{a}$ is perpendicular to $\mathbf{b}$.

10 Find a vector which is perpendicular to both $\mathbf{a}$ and $\mathbf{b}$, where:

a $\mathbf{a} = \mathbf{i} + \mathbf{j} - 3\mathbf{k}$, $\mathbf{b} = 5\mathbf{i} - 2\mathbf{j} - \mathbf{k}$

b $\mathbf{a} = 2\mathbf{i} + 3\mathbf{j} - 4\mathbf{k}$, $\mathbf{b} = \mathbf{i} - 6\mathbf{j} + 3\mathbf{k}$

c $\mathbf{a} = 4\mathbf{i} - 4\mathbf{j} - \mathbf{k}$, $\mathbf{b} = -2\mathbf{i} - 9\mathbf{j} + 6\mathbf{k}$

11 The points A and B have position vectors $2\mathbf{i} + 5\mathbf{j} + \mathbf{k}$ and $6\mathbf{i} + \mathbf{j} - 2\mathbf{k}$ respectively, and O is the origin.
Calculate each of the angles in $\triangle OAB$, giving your answers in degrees to 1 decimal place.

12 The points A, B and C have position vectors $\mathbf{i} + 3\mathbf{j} + \mathbf{k}$, $2\mathbf{i} + 7\mathbf{j} - 3\mathbf{k}$ and $4\mathbf{i} - 5\mathbf{j} + 2\mathbf{k}$ respectively.

a Find, as surds, the lengths of AB and BC.

b Calculate, in degrees to 1 decimal place, the size of $\angle ABC$.

13 Given that the points A and B have coordinates $(7, 4, 4)$ and $(2, -2, -1)$ respectively, use a vector method to find the value of $\cos AOB$, where O is the origin.

Prove that the area of $\triangle AOB$ is $\frac{5\sqrt{29}}{2}$.

14 AB is a diameter of a circle centred at the origin O, and P is any point on the circumference of the circle.
Using the position vectors of A, B and P, prove (using a scalar product) that AP is perpendicular to BP (i.e. the angle in the semicircle is a right angle).

15 Use a vector method to prove that the diagonals of the square $OABC$ cross at right angles.

5.8 You need to know how to write the equation of a straight line in vector form.

Suppose a straight line passes through a given point A, with position vector $\mathbf{a}$, and is parallel to the given vector $\mathbf{b}$. Only one such line is possible.

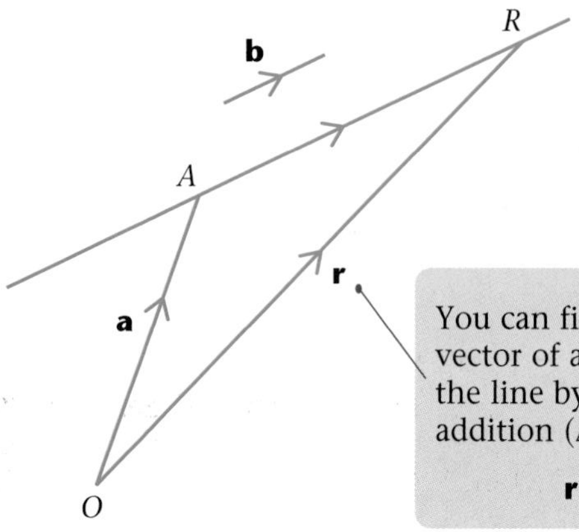

Since $\overrightarrow{AR}$ is parallel to $\mathbf{b}$, $\overrightarrow{AR} = t\mathbf{b}$, where t is a scalar.
The vector $\mathbf{b}$ is called the direction vector of the line.
So the position vector $\mathbf{r}$ can be written as $\mathbf{a} + t\mathbf{b}$.

You can find the position vector of any point R on the line by using vector addition ($\triangle OAR$):

$\mathbf{r} = \mathbf{a} + \overrightarrow{AR}$

■ **A vector equation of a straight line passing through the point A with position vector $\mathbf{a}$, and parallel to the vector $\mathbf{b}$, is**

$$\mathbf{r} = \mathbf{a} + t\mathbf{b}$$

where t is a scalar parameter.

By taking different values of the parameter t, you can find the position vectors of different points that lie on the straight line.

Example 29

Find a vector equation of the straight line which passes through the point A, with position vector $3\mathbf{i} - 5\mathbf{j} + 4\mathbf{k}$, and is parallel to the vector $7\mathbf{i} - 3\mathbf{k}$.

Here $\mathbf{a} = \begin{pmatrix} 3 \\ -5 \\ 4 \end{pmatrix}$ and $\mathbf{b} = \begin{pmatrix} 7 \\ 0 \\ -3 \end{pmatrix}$

b is the direction vector.

An equation of the line is

$$\mathbf{r} = \begin{pmatrix} 3 \\ -5 \\ 4 \end{pmatrix} + t\begin{pmatrix} 7 \\ 0 \\ -3 \end{pmatrix}$$

or $\quad \mathbf{r} = (3\mathbf{i} - 5\mathbf{j} + 4\mathbf{k}) + t(7\mathbf{i} - 3\mathbf{k})$

or $\quad \mathbf{r} = (3 + 7t)\mathbf{i} + (-5)\mathbf{j} + (4 - 3t)\mathbf{k}$

or $\quad \mathbf{r} = \begin{pmatrix} 3 + 7t \\ -5 \\ 4 - 3t \end{pmatrix}$

You sometimes need to show the separate x, y, z components in terms of t.

Now suppose a straight line passes through two given points C and D, with position vectors **c** and **d** respectively. Again, only one such line is possible.

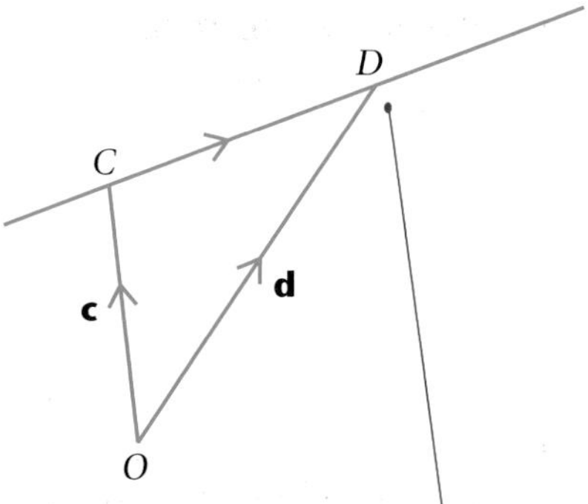

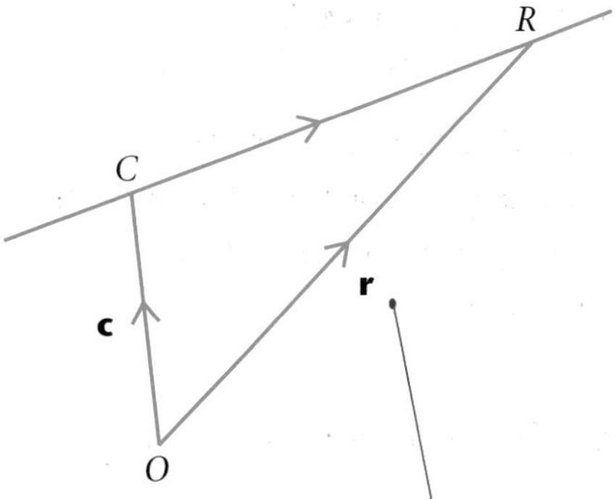

You can use CD as a direction vector for the line: $\overrightarrow{CD} = \mathbf{d} - \mathbf{c}$ (see Section 5.3).

You can now use one of the two given points and the direction vector to form an equation for the straight line.

■ **A vector equation of a straight line passing through the points *C* and *D*, with position vectors c and d respectively, is**

$\mathbf{r} = \mathbf{c} + t(\mathbf{d} - \mathbf{c})$

where *t* is a scalar parameter.

You could also have used the point *D*, giving $\mathbf{r} = \mathbf{d} + t(\mathbf{d} - \mathbf{c})$.

Example 30

Find a vector equation of the straight line which passes through the points *A* and *B*, with coordinates (4, 5, −1) and (6, 3, 2) respectively.

$$\mathbf{a} = \begin{pmatrix} 4 \\ 5 \\ -1 \end{pmatrix} \quad \mathbf{b} = \begin{pmatrix} 6 \\ 3 \\ 2 \end{pmatrix}$$

Write down the position vectors of *A* and *B*.

$$\mathbf{b} - \mathbf{a} = \begin{pmatrix} 6 \\ 3 \\ 2 \end{pmatrix} - \begin{pmatrix} 4 \\ 5 \\ -1 \end{pmatrix} = \begin{pmatrix} 2 \\ -2 \\ 3 \end{pmatrix}$$

Find a direction vector for the line.

$$\mathbf{r} = \begin{pmatrix} 4 \\ 5 \\ -1 \end{pmatrix} + t\begin{pmatrix} 2 \\ -2 \\ 3 \end{pmatrix}$$

Use one of the given points to form the equation.

The equation could be written in other ways:

$$\mathbf{r} = (4\mathbf{i} + 5\mathbf{j} - \mathbf{k}) + t(2\mathbf{i} - 2\mathbf{j} + 3\mathbf{k})$$

$$\mathbf{r} = (4 + 2t)\mathbf{i} + (5 - 2t)\mathbf{j} + (-1 + 3t)\mathbf{k}$$

$$\mathbf{r} = \begin{pmatrix} 4 + 2t \\ 5 - 2t \\ -1 + 3t \end{pmatrix}$$

Example 31

The straight line *l* has vector equation $\mathbf{r} = (3\mathbf{i} + 2\mathbf{j} - 5\mathbf{k}) + t(\mathbf{i} - 6\mathbf{j} - 2\mathbf{k})$.
Given that the point $(a, b, 0)$ lies on *l*, find the value of *a* and the value of *b*.

$$\mathbf{r} = \begin{pmatrix} 3 + t \\ 2 - 6t \\ -5 - 2t \end{pmatrix}$$

You can write the equation in this form.

$$-5 - 2t = 0$$

Use the *z*-coordinate (zero) to find the value of *t*.

$$t = -2\tfrac{1}{2}$$

$$a = 3 + t = \tfrac{1}{2}$$

$$b = 2 - 6t = 17$$

$$a = \tfrac{1}{2} \text{ and } b = 17$$

Find *a* and *b* using the value of *t*.

Example 32

The straight line l has vector equation $\mathbf{r} = (2\mathbf{i} + 5\mathbf{j} - 3\mathbf{k}) + t(6\mathbf{i} - 2\mathbf{j} + 4\mathbf{k})$.
Show that another vector equation of l is $\mathbf{r} = (8\mathbf{i} + 3\mathbf{j} + \mathbf{k}) + t(3\mathbf{i} - \mathbf{j} + 2\mathbf{k})$.

Since $\begin{pmatrix} 6 \\ -2 \\ 4 \end{pmatrix} = 2\begin{pmatrix} 3 \\ -1 \\ 2 \end{pmatrix}$, these two vectors are parallel.

So $\begin{pmatrix} 3 \\ -1 \\ 2 \end{pmatrix}$ can also be used as the direction vector.

So another form of the equation is l is

$$\mathbf{r} = \begin{pmatrix} 2 \\ 5 \\ -3 \end{pmatrix} + t\begin{pmatrix} 3 \\ -1 \\ 2 \end{pmatrix}.$$

If $t = 2$, $\mathbf{r} = \begin{pmatrix} 8 \\ 3 \\ 1 \end{pmatrix}$

You need to show that (8, 3, 1) lies on l. Look for a t-value that gives the position vector of this point.

So the point (8, 3, 1) also lies on l.

So another form of the equation of l is

$$\mathbf{r} = \begin{pmatrix} 8 \\ 3 \\ 1 \end{pmatrix} + t\begin{pmatrix} 3 \\ -1 \\ 2 \end{pmatrix}.$$

Exercise 5H

1 Find a vector equation of the straight line which passes through the point A, with position vector $\mathbf{a}$, and is parallel to the vector $\mathbf{b}$:

a $\mathbf{a} = 6\mathbf{i} + 5\mathbf{j} - \mathbf{k}$, $\mathbf{b} = 2\mathbf{i} - 3\mathbf{j} - \mathbf{k}$

b $\mathbf{a} = 2\mathbf{i} + 5\mathbf{j}$, $\mathbf{b} = \mathbf{i} + \mathbf{j} + \mathbf{k}$

c $\mathbf{a} = -7\mathbf{i} + 6\mathbf{j} + 2\mathbf{k}$, $\mathbf{b} = 3\mathbf{i} + \mathbf{j} + 2\mathbf{k}$

d $\mathbf{a} = \begin{pmatrix} 2 \\ 0 \\ 4 \end{pmatrix}$, $\mathbf{b} = \begin{pmatrix} -3 \\ 2 \\ 1 \end{pmatrix}$

e $\mathbf{a} = \begin{pmatrix} 6 \\ -11 \\ 2 \end{pmatrix}$, $\mathbf{b} = \begin{pmatrix} 0 \\ 5 \\ -2 \end{pmatrix}$

2 Calculate, to 1 decimal place, the distance between the point P, where $t = 1$, and the point Q, where $t = 5$, on the line with equation:

a $\mathbf{r} = (2\mathbf{i} - \mathbf{j} + \mathbf{k}) + t(3\mathbf{i} - 8\mathbf{j} - \mathbf{k})$

b $\mathbf{r} = (\mathbf{i} + 4\mathbf{j} + \mathbf{k}) + t(6\mathbf{i} - 2\mathbf{j} + 3\mathbf{k})$

c $\mathbf{r} = (2\mathbf{i} + 5\mathbf{k}) + t(-3\mathbf{i} + 4\mathbf{j} - \mathbf{k})$

3 Find a vector equation for the line which is parallel to the z-axis and passes through the point $(4, -3, 8)$.

4 Find a vector equation for the line which passes through the points:

a $(2, 1, 9)$ and $(4, -1, 8)$

b $(-3, 5, 0)$ and $(7, 2, 2)$

c $(1, 11, -4)$ and $(5, 9, 2)$

d $(-2, -3, -7)$ and $(12, 4, -3)$

5 The point $(1, p, q)$ lies on the line l. Find the values of p and q, given that the equation is l is:

a $\mathbf{r} = (2\mathbf{i} - 3\mathbf{j} + \mathbf{k}) + t(\mathbf{i} - 4\mathbf{j} - 9\mathbf{k})$

b $\mathbf{r} = (-4\mathbf{i} + 6\mathbf{j} - \mathbf{k}) + t(2\mathbf{i} - 5\mathbf{j} - 8\mathbf{k})$

c $\mathbf{r} = (16\mathbf{i} - 9\mathbf{j} - 10\mathbf{k}) + t(3\mathbf{i} + 2\mathbf{j} + \mathbf{k})$

5.9 You need to be able to determine whether two given straight lines intersect.

When you need to deal with more than one straight line in the same question, use a different parameter for each line.

The letters t and s are often used as parameters.
Greek letters λ and μ are also commonly used as parameters.

In three dimensions, two straight lines will not generally intersect. The next example, however, deals with two straight lines that do intersect, and shows you how to prove this.

Example 33

Show that the lines with vector equations

$$\mathbf{r} = (3\mathbf{i} + 8\mathbf{j} - 2\mathbf{k}) + t(2\mathbf{i} - \mathbf{j} + 3\mathbf{k})$$
$$\text{and } \mathbf{r} = (7\mathbf{i} + 4\mathbf{j} + 3\mathbf{k}) + s(2\mathbf{i} + \mathbf{j} + 4\mathbf{k})$$

intersect, and find the position vector of their point of intersection.

$$\mathbf{r} = \begin{pmatrix} 3+2t \\ 8-t \\ -2+3t \end{pmatrix} \quad \mathbf{r} = \begin{pmatrix} 7+2s \\ 4+s \\ 3+4s \end{pmatrix}$$

At an intersection point,

$$\begin{pmatrix} 3+2t \\ 8-t \\ -2+3t \end{pmatrix} = \begin{pmatrix} 7+2s \\ 4+s \\ 3+4s \end{pmatrix}$$

$$3+2t = 7+2s$$
$$8-t = 4+s$$

Equate the x components.
Equate the y components.

$$3+2t = 7+2s$$
$$16-2t = 8+2s$$
$$19 = 15+4s$$
$$s = 1$$
$$3+2t = 7+2$$
$$t = 3$$

Solve simultaneously.

If the lines intersect,

$-2+3t = 3+4s$ must be true.

z components must also be equal

$$-2+3t = -2+9 = 7$$
$$3+4s = 3+4 = 7$$

Check that $s = 1$, $t = 3$ gives equal z components.

The z components are also equal, so the lines do intersect.

The intersection point has position vector:

$$\begin{pmatrix} 3+2t \\ 8-t \\ -2+3t \end{pmatrix}$$

With $t = 3$: $\mathbf{r} = \begin{pmatrix} 9 \\ 5 \\ 7 \end{pmatrix}$ or $\mathbf{r} = 9\mathbf{i} + 5\mathbf{j} + 7\mathbf{k}$

Exercise 5I

In each question, determine whether the lines with the given equations intersect. If they do intersect, find the coordinates of their point of intersection.

1 $\mathbf{r} = \begin{pmatrix} 2 \\ 4 \\ -7 \end{pmatrix} + t\begin{pmatrix} 2 \\ 1 \\ 3 \end{pmatrix}$ and $\mathbf{r} = \begin{pmatrix} 1 \\ 14 \\ 16 \end{pmatrix} + s\begin{pmatrix} 1 \\ -1 \\ -2 \end{pmatrix}$

2 $\mathbf{r} = \begin{pmatrix} 2 \\ 2 \\ -3 \end{pmatrix} + t\begin{pmatrix} 9 \\ -2 \\ -1 \end{pmatrix}$ and $\mathbf{r} = \begin{pmatrix} 3 \\ -1 \\ 2 \end{pmatrix} + s\begin{pmatrix} 2 \\ -1 \\ 3 \end{pmatrix}$

3 $\mathbf{r} = \begin{pmatrix} 12 \\ 4 \\ -6 \end{pmatrix} + t\begin{pmatrix} -2 \\ 1 \\ 4 \end{pmatrix}$ and $\mathbf{r} = \begin{pmatrix} 8 \\ -2 \\ 6 \end{pmatrix} + s\begin{pmatrix} 2 \\ 1 \\ -5 \end{pmatrix}$

4 $\mathbf{r} = \begin{pmatrix} 1 \\ 0 \\ 4 \end{pmatrix} + t\begin{pmatrix} 4 \\ 2 \\ 6 \end{pmatrix}$ and $\mathbf{r} = \begin{pmatrix} -2 \\ -9 \\ 12 \end{pmatrix} + s\begin{pmatrix} 1 \\ 2 \\ -1 \end{pmatrix}$

5 $\mathbf{r} = \begin{pmatrix} 3 \\ -3 \\ 1 \end{pmatrix} + t\begin{pmatrix} 2 \\ 1 \\ -4 \end{pmatrix}$ and $\mathbf{r} = \begin{pmatrix} 3 \\ 4 \\ 2 \end{pmatrix} + s\begin{pmatrix} 6 \\ -4 \\ 1 \end{pmatrix}$

5.10 You need to be able to calculate the angle between two straight lines.

■ **The acute angle θ between two straight lines is given by**

$$\cos\theta = \left|\frac{\mathbf{a}.\mathbf{b}}{|\mathbf{a}||\mathbf{b}|}\right|$$

where a and b are direction vectors of the lines.

Example 34

Find, to 1 decimal place, the acute angle between the lines with vector equations

$$\mathbf{r} = (2\mathbf{i} + \mathbf{j} + \mathbf{k}) + t(3\mathbf{i} - 8\mathbf{j} - \mathbf{k})$$
and $$\mathbf{r} = (7\mathbf{i} + 4\mathbf{j} + \mathbf{k}) + s(2\mathbf{i} + 2\mathbf{j} + 3\mathbf{k})$$

$\mathbf{a} = \begin{pmatrix} 3 \\ -8 \\ -1 \end{pmatrix}$ and $\mathbf{b} = \begin{pmatrix} 2 \\ 2 \\ 3 \end{pmatrix}$ — Use the direction vectors.

$\cos\theta = \dfrac{\mathbf{a}.\mathbf{b}}{|\mathbf{a}||\mathbf{b}|}$ — Find the angle between the 2 vectors.

$\mathbf{a}.\mathbf{b} = \begin{pmatrix} 3 \\ -8 \\ -1 \end{pmatrix} . \begin{pmatrix} 2 \\ 2 \\ 3 \end{pmatrix}$

$= 6 - 16 - 3 = -13$

$|\mathbf{a}| = \sqrt{3^2 + (-8)^2 + (-1)^2} = \sqrt{74}$

$|\mathbf{b}| = \sqrt{2^2 + 2^2 + 3^2} = \sqrt{17}$

$\cos\theta = -\dfrac{13}{\sqrt{74}\sqrt{17}}$ — Use the formula for $\cos\theta$.

$\theta = 68.5°$ (1 d.p.) — This is the angle between 2 *vectors*.

So the acute angle between the lines is
$180° - 111.5° = 68.5°$ (1 d.p.)

Exercise 5J

In Questions 1 to 5, find, to 1 decimal place, the acute angle between the lines with the given vector equations:

1 $\mathbf{r} = (2\mathbf{i} + \mathbf{j} + \mathbf{k}) + t(3\mathbf{i} - 5\mathbf{j} - \mathbf{k})$
and $\mathbf{r} = (7\mathbf{i} + 4\mathbf{j} + \mathbf{k}) + s(2\mathbf{i} + \mathbf{j} - 9\mathbf{k})$

2 $\mathbf{r} = (\mathbf{i} - \mathbf{j} + 7\mathbf{k}) + t(-2\mathbf{i} - \mathbf{j} + 3\mathbf{k})$
and $\mathbf{r} = (8\mathbf{i} + 5\mathbf{j} - \mathbf{k}) + s(-4\mathbf{i} - 2\mathbf{j} + \mathbf{k})$

3 $\mathbf{r} = (3\mathbf{i} + 5\mathbf{j} - \mathbf{k}) + t(\mathbf{i} + \mathbf{j} + \mathbf{k})$
and $\mathbf{r} = (-\mathbf{i} + 11\mathbf{j} + 5\mathbf{k}) + s(2\mathbf{i} - 7\mathbf{j} + 3\mathbf{k})$

4 $\mathbf{r} = (\mathbf{i} + 6\mathbf{j} - \mathbf{k}) + t(8\mathbf{i} - \mathbf{j} - 2\mathbf{k})$
and $\mathbf{r} = (6\mathbf{i} + 9\mathbf{j}) + s(\mathbf{i} + 3\mathbf{j} - 7\mathbf{k})$

5 $\mathbf{r} = (2\mathbf{i} + \mathbf{k}) + t(11\mathbf{i} + 5\mathbf{j} - 3\mathbf{k})$
and $\mathbf{r} = (\mathbf{i} + \mathbf{j}) + s(-3\mathbf{i} + 5\mathbf{j} + 4\mathbf{k})$

6 The straight lines l_1 and l_2 have vector equations $\mathbf{r} = (\mathbf{i} + 4\mathbf{j} + 2\mathbf{k}) + t(8\mathbf{i} + 5\mathbf{j} + \mathbf{k})$ and $\mathbf{r} = (\mathbf{i} + 4\mathbf{j} + 2\mathbf{k}) + s(3\mathbf{i} + \mathbf{j})$ respectively, and P is the point with coordinates (1, 4, 2).

a Show that the point $Q(9, 9, 3)$ lies on l_1.

b Find the cosine of the acute angle between l_1 and l_2.

c Find the possible coordinates of the point R, such that R lies on l_2 and $PQ = PR$.

Mixed exercise 5K

1 With respect to an origin O, the position vectors of the points L, M and N are $(4\mathbf{i} + 7\mathbf{j} + 7\mathbf{k})$, $(\mathbf{i} + 3\mathbf{j} + 2\mathbf{k})$ and $(2\mathbf{i} + 4\mathbf{j} + 6\mathbf{k})$ respectively.

a Find the vectors $\overrightarrow{ML}$ and $\overrightarrow{MN}$.

b Prove that $\cos \angle LMN = \frac{9}{10}$. **E**

2 The position vectors of the points A and B relative to an origin O are $5\mathbf{i} + 4\mathbf{j} + \mathbf{k}$, $-\mathbf{i} + \mathbf{j} - 2\mathbf{k}$ respectively. Find the position vector of the point P which lies on AB produced such that $AP = 2BP$. **E**

3 Points A, B, C, D in a plane have position vectors $\mathbf{a} = 6\mathbf{i} + 8\mathbf{j}$, $\mathbf{b} = \frac{3}{2}\mathbf{a}$, $\mathbf{c} = 6\mathbf{i} + 3\mathbf{j}$, $\mathbf{d} = \frac{5}{3}\mathbf{c}$ respectively. Write down vector equations of the lines AD and BC and find the position vector of their point of intersection. **E**

4 Find the point of intersection of the line through the points (2, 0, 1) and (−1, 3, 4) and the line through the points (−1, 3, 0) and (4, −2, 5). Calculate the acute angle between the two lines. **E**

5 Show that the lines

$$\mathbf{r} = (-2\mathbf{i} + 5\mathbf{j} - 11\mathbf{k}) + \lambda(3\mathbf{i} + \mathbf{j} + 3\mathbf{k})$$
$$\mathbf{r} = 8\mathbf{i} + 9\mathbf{j} + \mu(4\mathbf{i} + 2\mathbf{j} + 5\mathbf{k})$$

intersect. Find the position vector of their common point. E

6 Find a vector that is perpendicular to both $2\mathbf{i} + \mathbf{j} - \mathbf{k}$ and $\mathbf{i} + \mathbf{j} - 2\mathbf{k}$. E

7 State a vector equation of the line passing through the points A and B whose position vectors are $\mathbf{i} - \mathbf{j} + 3\mathbf{k}$ and $\mathbf{i} + 2\mathbf{j} + 2\mathbf{k}$ respectively. Determine the position vector of the point C which divides the line segment AB internally such that $AC = 2CB$. E

8 Vectors $\mathbf{r}$ and $\mathbf{s}$ are given by

$$\mathbf{r} = \lambda\mathbf{i} + (2\lambda - 1)\mathbf{j} - \mathbf{k}$$
$$\mathbf{s} = (1 - \lambda)\mathbf{i} + 3\lambda\mathbf{j} + (4\lambda - 1)\mathbf{k}$$

where λ is a scalar.

a Find the values of λ for which $\mathbf{r}$ and $\mathbf{s}$ are perpendicular.

When $\lambda = 2$, $\mathbf{r}$ and $\mathbf{s}$ are the position vectors of the points A and B respectively, referred to an origin O.

b Find $\overrightarrow{AB}$.

c Use a scalar product to find the size of $\angle BAO$, giving your answer to the nearest degree. E

9 With respect to an origin O, the position vectors of the points L and M are $2\mathbf{i} - 3\mathbf{j} + 3\mathbf{k}$ and $5\mathbf{i} + \mathbf{j} + c\mathbf{k}$ respectively, where c is a constant. The point N is such that $OLMN$ is a rectangle.

a Find the value of c.

b Write down the position vector of N.

c Find, in the form $\mathbf{r} = \mathbf{p} + t\mathbf{q}$, an equation of the line MN. E

10 The point A has coordinates $(7, -1, 3)$ and the point B has coordinates $(10, -2, 2)$. The line l has vector equation $\mathbf{r} = \mathbf{i} + \mathbf{j} + \mathbf{k} + \lambda(3\mathbf{i} - \mathbf{j} + \mathbf{k})$, where λ is a real parameter.

a Show that the point A lies on the line l.

b Find the length of AB.

c Find the size of the acute angle between the line l and the line segment AB, giving your answer to the nearest degree.

d Hence, or otherwise, calculate the perpendicular distance from B to the line l, giving your answer to two significant figures. E

11 Referred to a fixed origin O, the points A and B have position vectors $(5\mathbf{i} - \mathbf{j} - \mathbf{k})$ and $(\mathbf{i} - 5\mathbf{j} + 7\mathbf{k})$ respectively.

a Find an equation of the line AB.

b Show that the point C with position vector $4\mathbf{i} - 2\mathbf{j} + \mathbf{k}$ lies on AB.

c Show that OC is perpendicular to AB.

d Find the position vector of the point D, where $D \neq A$, on AB such that $|\overrightarrow{OD}| = |\overrightarrow{OA}|$.

E

12 Referred to a fixed origin O, the points A, B and C have position vectors $(9\mathbf{i} - 2\mathbf{j} + \mathbf{k})$, $(6\mathbf{i} + 2\mathbf{j} + 6\mathbf{k})$ and $(3\mathbf{i} + p\mathbf{j} + q\mathbf{k})$ respectively, where p and q are constants.

a Find, in vector form, an equation of the line l which passes through A and B.

Given that C lies on l:

b Find the value of p and the value of q.

c Calculate, in degrees, the acute angle between OC and AB,

The point D lies on AB and is such that OD is perpendicular to AB.

d Find the position vector of D.

E

13 Referred to a fixed origin O, the points A and B have position vectors $(\mathbf{i} + 2\mathbf{j} - 3\mathbf{k})$ and $(5\mathbf{i} - 3\mathbf{j})$ respectively.

a Find, in vector form, an equation of the line l_1 which passes through A and B.

The line l_2 has equation $\mathbf{r} = (4\mathbf{i} - 4\mathbf{j} + 3\mathbf{k}) + \mu(\mathbf{i} - 2\mathbf{j} + 2\mathbf{k})$, where μ is a scalar parameter.

b Show that A lies on l_2.

c Find, in degrees, the acute angle between the lines l_1 and l_2.

The point C with position vector $(2\mathbf{i} - \mathbf{k})$ lies on l_2.

d Find the shortest distance from C to the line l_1.

E

14 Two submarines are travelling in straight lines through the ocean. Relative to a fixed origin, the vector equations of the two lines, l_1 and l_2, along which they travel are

$$\mathbf{r} = 3\mathbf{i} + 4\mathbf{j} - 5\mathbf{k} + \lambda(\mathbf{i} - 2\mathbf{j} + 2\mathbf{k})$$
$$\text{and } \mathbf{r} = 9\mathbf{i} + \mathbf{j} - 2\mathbf{k} + \mu(4\mathbf{i} + \mathbf{j} - \mathbf{k})$$

where λ and μ are scalars.

a Show that the submarines are moving in perpendicular directions.

b Given that l_1 and l_2 intersect at the point A, find the position vector of A. The point B has position vector $10\mathbf{j} - 11\mathbf{k}$.

c Show that only one of the submarines passes through the point B.

d Given that 1 unit on each coordinate axis represents 100 m, find, in km, the distance AB.

E

Summary of key points

1 A vector is a quantity that has both magnitude and direction.

2 Vectors that are equal have both the same magnitude and the same direction.

3 Two vectors are added using the 'triangle law'.

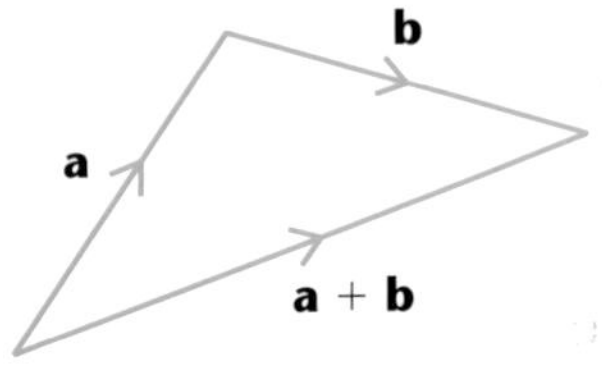

4 Adding the vectors $\overrightarrow{PQ}$ and $\overrightarrow{QP}$ gives the zero vector **0**.
($\overrightarrow{PQ} + \overrightarrow{QP} = \mathbf{0}$)

5 The modulus of a vector is another name for its magnitude.
- The modulus of the vector **a** is written as $|\mathbf{a}|$.
- The modulus of the vector $\overrightarrow{PQ}$ is written as $|\overrightarrow{PQ}|$.

6 The vector $-\mathbf{a}$ has the same magnitude as the vector **a** but is in the opposite direction.

7 Any vector parallel to the vector **a** may be written as $\lambda\mathbf{a}$, where λ is a non-zero scalar.

8 $\mathbf{a} - \mathbf{b}$ is defined to be $\mathbf{a} + (-\mathbf{b})$.

9 A unit vector is a vector which has magnitude (or modulus) 1 unit.

10 If $\lambda\mathbf{a} + \mu\mathbf{b} = \alpha\mathbf{a} + \beta\mathbf{b}$, and the non-zero vectors **a** and **b** are not parallel, then $\lambda = \alpha$ and $\mu = \beta$.

11 The position vector of a point A is the vector $\overrightarrow{OA}$, where O is the origin. $\overrightarrow{OA}$ is usually written as vector **a**.

12 $\overrightarrow{AB} = \mathbf{b} - \mathbf{a}$, where **a** and **b** are the position vectors of A and B respectively.

13 The vectors **i**, **j** and **k** are unit vectors parallel to the x-axis, the y-axis and the z-axis and in the direction of x increasing, y increasing and z increasing, respectively.

14 The modulus (or magnitude) of $x\mathbf{i} + y\mathbf{j}$ is $\sqrt{x^2 + y^2}$.

15 The vector $x\mathbf{i} + y\mathbf{j} + z\mathbf{k}$ may be written as a column matrix $\begin{pmatrix} x \\ y \\ z \end{pmatrix}$.

16 The distance from the origin to the point (x, y, z) is $\sqrt{x^2 + y^2 + z^2}$.

17 The distance between the points (x_1, y_1, z_1) and (x_2, y_2, z_2) is $\sqrt{(x_1 - x_2)^2 + (y_1 - y_2)^2 + (z_1 - z_2)^2}$.

18 The modulus (or magnitude) of $x\mathbf{i} + y\mathbf{j} + z\mathbf{k}$ is $\sqrt{x^2 + y^2 + z^2}$.

19 The scalar product of two vectors **a** and **b** is written as **a.b**, and defined by

$$\mathbf{a.b} = |\mathbf{a}|\,|\mathbf{b}| \cos\theta,$$

where θ is the angle between **a** and **b**.

20 If **a** and **b** are the position vectors of the points A and B, then

$$\cos AOB = \frac{\mathbf{a.b}}{|\mathbf{a}|\,|\mathbf{b}|}$$

21 The non-zero vectors **a** and **b** are perpendicular if and only if $\mathbf{a.b} = 0$.

22 If **a** and **b** are parallel, $\mathbf{a.b} = |\mathbf{a}|\,|\mathbf{b}|$.

- In particular, $\mathbf{a.a} = |\mathbf{a}|^2$.

23 If $\mathbf{a} = a_1\mathbf{i} + a_2\mathbf{j} + a_3\mathbf{k}$ and $\mathbf{b} = b_1\mathbf{i} + b_2\mathbf{j} + b_3\mathbf{k}$

$$\mathbf{a.b} = \begin{pmatrix} a_1 \\ a_2 \\ a_3 \end{pmatrix} . \begin{pmatrix} b_1 \\ b_2 \\ b_3 \end{pmatrix} = a_1b_1 + a_2b_2 + a_3b_3$$

24 A vector equation of a straight line passing through the point A with position vector **a**, and parallel to the vector **b**, is

$$\mathbf{r} = \mathbf{a} + t\mathbf{b}$$

where t is a scalar parameter.

25 A vector equation of a straight line passing through the points C and D, with position vectors **c** and **d** respectively, is

$$\mathbf{r} = \mathbf{c} + t(\mathbf{d} - \mathbf{c})$$

where t is a scalar parameter.

26 The acute angle θ between two straight lines is given by

$$\cos\theta = \left|\frac{\mathbf{a.b}}{|\mathbf{a}|\,|\mathbf{b}|}\right|$$

where **a** and **b** are direction vectors of the lines.

6 Integration

In this chapter you will learn how to integrate more complicated functions.

6.1 You need to be able to integrate standard functions.

You met the first result in the list below in your C1 book. The others are the reverse of ones you have already met in Chapter 8 of C3.

■ **You should be familiar with the following integrals:**

① $\int x^n = \frac{x^{n+1}}{n+1} + C$

② $\int e^x = e^x + C$

③ $\int \frac{1}{x} = \ln|x| + C$

④ $\int \cos x = \sin x + C$

⑤ $\int \sin x = -\cos x + C$

⑥ $\int \sec^2 x = \tan x + C$

⑦ $\int \operatorname{cosec} x \cot x = -\operatorname{cosec} x + C$

⑧ $\int \operatorname{cosec}^2 x = -\cot x + C$

⑨ $\int \sec x \tan x = \sec x + C$

Hint: When finding $\int \frac{1}{x}\,dx$ it is usual to write the answer as $\ln|x|$. The modulus sign removes difficulties that may raise when evaluating the integral. This will be explained in more detail in Section 6.5 but this form will be used throughout this chapter.

Example 1

Find the following integrals:

a $\int\left(2\cos x + \frac{3}{x} - \sqrt{x}\right)dx$ **b** $\int\left(\frac{\cos x}{\sin^2 x} - 2e^x\right)dx$

a $\int 2\cos x\,dx = 2\sin x$

Integrate each term separately.

Use ④.

$\int \frac{3}{x}\,dx = 3\ln|x|$

Use ③.

$\int \sqrt{x}\,dx = \int x^{\frac{1}{2}}\,dx = \frac{2}{3}x^{\frac{3}{2}}$

Use ①.

So $\int\left(2\cos x + \frac{3}{x} - \sqrt{x}\right)dx$

$= 2\sin x + 3\ln|x| - \frac{2}{3}x^{\frac{3}{2}} + C$

This is an indefinite integral so don't forget the +C.

b $\dfrac{\cos x}{\sin^2 x} = \dfrac{\cos x}{\sin x}\dfrac{1}{\sin x} = \cot x \operatorname{cosec} x$

$\int(\cot x \operatorname{cosec} x)\,dx = -\operatorname{cosec} x$

$\int 2e^x\,dx = 2e^x$

So $\int\left(\dfrac{\cos x}{\sin^2 x} - 2e^x\right)dx$

$= -\operatorname{cosec} x - 2e^x + C$

Look at the list of integrals of standard functions and express the integrand in terms of these standard functions.

Remember the minus sign.

Exercise 6A

1 Integrate the following with respect to x:

a $3\sec^2 x + \dfrac{5}{x} + \dfrac{2}{x^2}$

b $5e^x - 4\sin x + 2x^3$

c $2(\sin x - \cos x + x)$

d $3\sec x \tan x - \dfrac{2}{x}$

e $5e^x + 4\cos x - \dfrac{2}{x^2}$

f $\dfrac{1}{2x} + 2\operatorname{cosec}^2 x$

g $\dfrac{1}{x} + \dfrac{1}{x^2} + \dfrac{1}{x^3}$

h $e^x + \sin x + \cos x$

i $2\operatorname{cosec} x \cot x - \sec^2 x$

j $e^x + \dfrac{1}{x} - \operatorname{cosec}^2 x$

2 Find the following integrals:

a $\int\left(\dfrac{1}{\cos^2 x} + \dfrac{1}{x^2}\right)dx$

b $\int\left(\dfrac{\sin x}{\cos^2 x} + 2e^x\right)dx$

c $\int\left(\dfrac{1+\cos x}{\sin^2 x} + \dfrac{1+x}{x^2}\right)dx$

d $\int\left(\dfrac{1}{\sin^2 x} + \dfrac{1}{x}\right)dx$

e $\int \sin x(1 + \sec^2 x)\,dx$

f $\int \cos x(1 + \operatorname{cosec}^2 x)\,dx$

g $\int \operatorname{cosec}^2 x(1 + \tan^2 x)\,dx$

h $\int \sec^2 x(1 - \cot^2 x)\,dx$

i $\int \sec^2 x(1 + e^x\cos^2 x)\,dx$

j $\int\left(\dfrac{1+\sin x}{\cos^2 x} + \cos^2 x \sec x\right)dx$

6.2 You can integrate some functions using the reverse of the chain rule.

This technique only works for linear transformations of functions such as $f(ax + b)$.

Example 2

Find the following integrals:

a $\int \cos(2x + 3)\,dx$ **b** $\int e^{4x+1}\,dx$ **c** $\int \sec^2 3x\,dx$

a Consider $y = \sin(2x + 3)$

Recall ④. So integrating a 'cos' function gives a 'sin' function.

So $\frac{dy}{dx} = \cos(2x + 3) \times 2$

Let $y = \sin(2x + 3)$ and differentiate using the chain rule.

Remember the 2 from the chain rule.

So $\int \cos(2x + 3)\,dx = \frac{1}{2}\sin(2x + 3) + C$

This is $2\times$ the required expression so you divide the $\sin(2x + 3)$ by 2.

b Consider $y = e^{4x+1}$

Recall ②. So integrating an 'exp' function gives an 'exp' function.

So $\frac{dy}{dx} = e^{4x+1} \times 4$

Let $y = e^{4x+1}$ and differentiate using the chain rule.

Remember the 4 comes from the chain rule.

So $\int e^{4x+1}\,dx = \frac{1}{4}e^{4x+1} + C$

This answer is 4 times the required expression so you divide by 4.

c Consider $y = \tan 3x$

Recall ⑥. Let $y = \tan 3x$ and differentiate using the chain rule.

So $\frac{dy}{dx} = \sec^2 3x \times 3$

This is 3 times the required expression so you divide by 3.

So $\int \sec^2 3x\,dx = \frac{1}{3}\tan 3x + C$

Example 2 illustrates the following general rule:

- $\int f'(ax + b)\,dx = \frac{1}{a}f(ax + b) + C$

You can generalise the list in Section 6.1 to give:

(10) $\int (ax+b)^n \, dx = \frac{1}{a} \frac{(ax+b)^{n+1}}{n+1} + C$

(11) $\int e^{ax+b} \, dx = \frac{1}{a} e^{ax+b} + C$

(12) $\int \frac{1}{ax+b} \, dx = \frac{1}{a} \ln|ax+b| + C$

(13) $\int \cos(ax+b) \, dx = \frac{1}{a} \sin(ax+b) + C$

(14) $\int \sin(ax+b) \, dx = -\frac{1}{a} \cos(ax+b) + C$

(15) $\int \sec^2(ax+b) \, dx = \frac{1}{a} \tan(ax+b) + C$

(16) $\int \operatorname{cosec}(ax+b)\cot(ax+b) \, dx = -\frac{1}{a} \operatorname{cosec}(ax+b) + C$

(17) $\int \operatorname{cosec}^2(ax+b) \, dx = -\frac{1}{a} \cot(ax+b) + C$

(18) $\int \sec(ax+b)\tan(ax+b) \, dx = \frac{1}{a} \sec(ax+b) + C$

In C4 it is probably best to learn *how* to work out these results using the chain rule, rather than trying to remember lots of formulae.

Example 3

Find the following integrals:

a $\int \frac{1}{3x+2} \, dx$ **b** $\int (2x+3)^4 \, dx$

a $\int \frac{1}{3x+2} \, dx = \frac{1}{3} \ln|3x+2| + C$ — Use (12).

b $\int (2x+3)^4 \, dx = \frac{1}{10}(2x+3)^5 + C$ — Use (10).

You can only use the results in the list above for linear transformations of functions. Integrals of the form $\int \cos(2x^2+3) \, dx$ do not give an answer like $\frac{1}{4x}\sin(2x^2+3)$ since differentiating this expression would require the quotient rule and would not give $\cos(2x^2+3)$. Expressions similar to this will be investigated in Section 6.5.

Exercise 6B

1 Integrate the following:

a $\sin(2x + 1)$ **b** $3e^{2x}$ **c** $4e^{x+5}$

d $\cos(1 - 2x)$ **e** $\text{cosec}^2 3x$ **f** $\sec 4x \tan 4x$

g $3\sin(\frac{1}{2}x + 1)$ **h** $\sec^2(2 - x)$ **i** $\text{cosec } 2x \cot 2x$

j $\cos 3x - \sin 3x$

2 Find the following integrals:

a $\int(e^{2x} - \frac{1}{2}\sin(2x - 1))\,dx$ **b** $\int(e^x + 1)^2\,dx$

c $\int \sec^2 2x(1 + \sin 2x)\,dx$ **d** $\int\left(\frac{3 - 2\cos(\frac{1}{2}x)}{\sin^2(\frac{1}{2}x)}\right)dx$

e $\int[e^{3-x} + \sin(3 - x) + \cos(3 - x)]\,dx$

3 Integrate the following:

a $\frac{1}{2x + 1}$ **b** $\frac{1}{(2x + 1)^2}$ **c** $(2x + 1)^2$ **d** $\frac{3}{4x - 1}$

e $\frac{3}{1 - 4x}$ **f** $\frac{3}{(1 - 4x)^2}$ **g** $(3x + 2)^5$ **h** $\frac{3}{(1 - 2x)^3}$

i $\frac{6}{(3 - 2x)^4}$ **j** $\frac{5}{3 - 2x}$

4 Find the following integrals

a $\int\left(3\sin(2x + 1) + \frac{4}{2x + 1}\right)dx$ **b** $\int[e^{5x} + (1 - x)^5]\,dx$

c $\int\left(\frac{1}{\sin^2 2x} + \frac{1}{1 + 2x} + \frac{1}{(1 + 2x)^2}\right)dx$ **d** $\int\left[(3x + 2)^2 + \frac{1}{(3x + 2)^2}\right]dx$

6.3 You can use trigonometric identities in integration.

Before you can integrate some trigonometric expressions you may need to replace the original expression with a function you can integrate from the lists on pages 82 and 85. You can do this using trigonometric identities.

Example 4

Find $\int \tan^2 x \, dx$.

Since $\quad \sec^2 x \equiv 1 + \tan^2 x$

Then $\quad \tan^2 x \equiv \sec^2 x - 1$

So $\int \tan^2 x \, dx = \int (\sec^2 x - 1) \, dx$

$= \int \sec^2 x \, dx - \int 1 \, dx$

$= \tan x - x + C$

You cannot integrate $\tan^2 x$ but you can integrate $\sec^2 x$ using ⑥ in the list on page 82.

Use ⑥.

In Book C3 you met identities for $\cos 2x$ in terms of $\sin^2 x$ and $\cos^2 x$. To integrate $\sin^2 x$ or $\cos^2 x$ you need to use one of these identities.

Example 5

Find $\int \sin^2 x \, dx$.

Recall $\quad \cos 2x \equiv 1 - 2\sin^2 x$

So $\quad \sin^2 x \equiv \frac{1}{2}(1 - \cos 2x)$

So $\int \sin^2 x \, dx = \int (\frac{1}{2} - \frac{1}{2}\cos 2x) \, dx$

$= \frac{1}{2}x - \frac{1}{4}\sin 2x + C$

You cannot integrate $\sin^2 x$ but you can write this in terms of $\cos 2x$.

Remember when integrating $\cos 2x$ you get an extra $\frac{1}{2}$. Use ⑬.

Example 6

Find:

a $\int \sin 3x \cos 3x \, dx$ **b** $\int (\sec x + \tan x)^2 \, dx$ **c** $\int \sin 3x \cos 2x \, dx$

a

$\int \sin 3x \cos 3x \, dx = \int \frac{1}{2}\sin 6x \, dx$

$= -\frac{1}{2} \times \frac{1}{6}\cos 6x + C$

$= -\frac{1}{12}\cos 6x + C$

Remember $\sin 2A = 2\sin A\cos A$, so $\sin 6x = 2\sin 3x \cos 3x$.

Use ⑭.

Simplify $\frac{1}{2} \times \frac{1}{6}$ to $\frac{1}{12}$.

b

$(\sec x + \tan x)^2$

$= \sec^2 x + 2\sec x \tan x + \tan^2 x$ — Multiply out the bracket.

$= \sec^2 x + 2\sec x \tan x + (\sec^2 x - 1)$ — Write $\tan^2 x$ as $\sec^2 x - 1$. Then all the terms are standard integrals.

$= 2\sec^2 x + 2\sec x \tan x - 1$

So $\int (\sec x + \tan x)^2 \, dx$

$= \int (2\sec^2 x + 2\sec x \tan x - 1) \, dx$ — Integrate each term using ⑥ and ⑨.

$= 2\tan x + 2\sec x - x + C$

c

Remember $\sin(A \pm B) = \sin A \cos B \pm \cos A \sin B$. So you need to use $A = 3x$ and $B = 2x$ to get a $\sin 3x \cos 2x$ term.

$\sin(3x + 2x) = \sin 3x \cos 2x + \cos 3x \sin 2x$

$\sin(3x - 2x) = \sin 3x \cos 2x - \cos 3x \sin 2x$

Adding gives

$\sin 5x + \sin x = 2\sin 3x \cos 2x$

So $\int \sin 3x \cos 2x \, dx$

$= \int \frac{1}{2}(\sin 5x + \sin x) \, dx$

$= \frac{1}{2}(-\frac{1}{5}\cos 5x - \cos x) + C$

$= -\frac{1}{10}\cos 5x - \frac{1}{2}\cos x + C$

Exercise 6C

1 Integrate the following:

a $\cot^2 x$

b $\cos^2 x$

c $\sin 2x \cos 2x$

d $(1 + \sin x)^2$

e $\tan^2 3x$

f $(\cot x - \operatorname{cosec} x)^2$

g $(\sin x + \cos x)^2$

h $\sin^2 x \cos^2 x$

i $\dfrac{1}{\sin^2 x \cos^2 x}$

j $(\cos 2x - 1)^2$

2 Find the following integrals:

a $\int\left(\frac{1-\sin x}{\cos^2 x}\right)dx$ **b** $\int\left(\frac{1+\cos x}{\sin^2 x}\right)dx$ **c** $\int\frac{\cos 2x}{\cos^2 x}\,dx$

d $\int\frac{\cos^2 x}{\sin^2 x}\,dx$ **e** $\int\frac{(1+\cos x)^2}{\sin^2 x}\,dx$ **f** $\int\frac{(1+\sin x)^2}{\cos^2 x}\,dx$

g $\int(\cot x-\tan x)^2\,dx$ **h** $\int(\cos x-\sin x)^2\,dx$ **i** $\int(\cos x-\sec x)^2\,dx$

j $\int\frac{\cos 2x}{1-\cos^2 2x}\,dx$

3 Find the following integrals:

a $\int\cos 2x\cos x\,dx$ **b** $\int 2\sin 5x\cos 3x\,dx$ **c** $\int 2\sin 3x\cos 5x\,dx$

d $\int 2\sin 2x\sin 5x\,dx$ **e** $4\int\cos 3x\cos 7x\,dx$ **f** $\int 2\cos 4x\cos 4x\,dx$

g $\int 2\cos 4x\sin 4x\,dx$ **h** $\int 2\sin 4x\sin 4x\,dx$

6.4 You can use partial fractions to integrate expressions.

In Chapter 1 you saw how to split certain rational expressions into partial fractions. This process is helpful in integration.

Example 7

Use partial fractions to find the following integrals:

a $\int\frac{x-5}{(x+1)(x-2)}\,dx$ **b** $\int\frac{8x^2-19x+1}{(2x+1)(x-2)^2}\,dx$ **c** $\int\frac{2}{(1-x^2)}\,dx$

a $\frac{x-5}{(x+1)(x-2)}\equiv\frac{A}{x+1}+\frac{B}{x-2}$

Split the expression to be integrated into partial fractions.

So $x-5\equiv A(x-2)+B(x+1)$

Put over the same denominator and compare numerators.

Let $x=-1$: $-6=A(-3)$ so $A=2$

Let $x=2$: $-3=B(3)$ so $B=-1$

Let $x=-1$ and 2.

So $\int\frac{x-5}{(x+1)(x-2)}\,dx$

$=\int\left(\frac{2}{x+1}-\frac{1}{x-2}\right)dx$

Rewrite the integral and integrate each term as in Section 6.2.

$=2\ln|x+1|-\ln|x-2|+C$

Remember to use the modulus when using ln in integration.

$=\ln\left|\frac{(x+1)^2}{x-2}\right|+C$

The answer could be left in this form, but sometimes you may be asked to combine the ln terms using the rules of logarithms met in Book C2.

b Let $I = \int \frac{8x^2 - 19x + 1}{(2x+1)(x-2)^2}\,dx$

It is sometimes useful to label the integral as I.

$$\frac{8x^2 - 19x + 1}{(2x+1)(x-2)^2} \equiv \frac{A}{2x+1} + \frac{B}{(x-2)^2} + \frac{C}{x-2}$$

Remember the partial fraction form for a repeated factor on the denominator.

$$8x^2 - 19x + 1 \equiv A(x-2)^2 + B(2x+1) + C(2x+1)(x-2)$$

Put over same denominator and compare numerators, then find the values of A, B and C.

Let $x = 2$

Then $-5 = 0 + 5B + 0$ so $B = -1$

Let $x = -\frac{1}{2}$

Then $12\frac{1}{2} = \frac{25}{4}A + 0 + 0$ so $A = 2$

Let $x = 0$ Then $1 = 4A - 2C + B$

So $1 = 8 - 2C - 1$ so $C = 3$

$$I = \int \left(\frac{2}{2x+1} - \frac{1}{(x-2)^2} + \frac{3}{x-2}\right) dx$$

Rewrite the integral using the partial fractions. Note that using I saves copying the question again.

$$I = \tfrac{2}{2}\ln|2x+1| + \frac{1}{x-2} + 3\ln|x-2| + C$$

Don't forget to divide by 2 when integrating $\frac{1}{2x+1}$ and remember that the integral of $\frac{1}{(x-2)^2}$ does not involve ln.

$$I = \ln|2x+1| + \frac{1}{x-2} + \ln|x-2|^3 + C$$

$$I = \ln|(2x+1)(x-2)^3| + \frac{1}{x-2} + C$$

Simplify using the laws of logarithms.

c Let $I = \int \frac{2}{(1-x^2)}\,dx$

$$\frac{2}{(1-x^2)} = \frac{2}{(1-x)(1+x)} = \frac{A}{1-x} + \frac{B}{1+x}$$

Remember that $(1 - x^2)$ can be factorised using the difference of two squares.

$2 = A(1+x) + B(1-x)$

$x = -1$ gives $2 = 2B$ so $B = 1$

$x = 1$ gives $2 = 2A$ so $A = 1$

So $I = \int \left(\frac{1}{1+x} + \frac{1}{1-x}\right) dx$ *

Rewrite the integral using the partial fractions.

$I = \ln|1+x| - \ln|1-x| + C$

or $I = \ln\left|\frac{1+x}{1-x}\right| + C$

Notice the minus sign that comes from integrating $\frac{1}{1-x}$.

You should notice that because the modulus sign is used in writing this integral, the answer could be

$$I = \ln\left|\frac{1+x}{x-1}\right| + C$$

since $|1 - x| = |x - 1|$. This can be found from line ✻ in the above example.

$$I = \int\left(\frac{1}{1+x} + \frac{1}{1-x}\right)\mathrm{d}x$$

$$I = \int\left(\frac{1}{1+x} - \frac{1}{x-1}\right)\mathrm{d}x$$

Since $\dfrac{1}{1-x} = -\dfrac{1}{x-1}$.

So $I = \ln|1+x| - \ln|x-1| + C$

So $I = \ln\left|\dfrac{1+x}{x-1}\right| + C$

Notice that no extra minus sign is needed when integrating $\dfrac{1}{x-1}$.

This use of the modulus sign, mentioned in Section 6.1, means that both cases can be incorporated in the one expression and this is one reason why the convention is used.

To integrate an improper fraction, you need to divide the numerator by the denominator.

Example 8

Find $\displaystyle\int \frac{9x^2 - 3x + 2}{9x^2 - 4}\,\mathrm{d}x$.

Let $I = \displaystyle\int \frac{9x^2 - 3x + 2}{9x^2 - 4}\,dx$ — First divide the numerator by $9x^2 - 4$.

$$\begin{array}{r} 1 \\ (9x^2 - 4)\;\overline{)9x^2 - 3x + 2} \\ \underline{9x^2 \qquad -4} \\ -3x + 6 \end{array}$$

$9x^2 \div 9x^2$ gives 1, so put this on top and subtract $9x^2 - 4$. This leaves a remainder of $-3x + 6$.

so $I = \displaystyle\int\left(1 + \frac{6 - 3x}{9x^2 - 4}\right)dx$

$\dfrac{6 - 3x}{9x^2 - 4} = \dfrac{A}{3x - 2} + \dfrac{B}{3x + 2}$ — Factorise $9x^2 - 4$ and then split into partial fractions.

$x = -\frac{2}{3} \Rightarrow 8 = -4B$ so $B = -2$

$x = \frac{2}{3} \Rightarrow 4 = 4A$ so $A = 1$

So $I = \displaystyle\int\left(1 + \frac{1}{3x - 2} - \frac{2}{3x + 2}\right)dx$ — Rewrite the integral using the partial fractions.

So $I = x + \frac{1}{3}\ln|3x - 2| - \frac{2}{3}\ln|3x + 2| + C$ — Integrate and don't forget the $\frac{1}{3}$.

Or $I = x + \frac{1}{3}\ln\left|\dfrac{3x - 2}{(3x + 2)^2}\right| + C$ — Simplify using the laws of logarithms.

Exercise 6D

1 Use partial fractions to integrate the following:

a $\dfrac{3x+5}{(x+1)(x+2)}$ **b** $\dfrac{3x-1}{(2x+1)(x-2)}$ **c** $\dfrac{2x-6}{(x+3)(x-1)}$ **d** $\dfrac{3}{(2+x)(1-x)}$

e $\dfrac{4}{(2x+1)(1-2x)}$ **f** $\dfrac{3(x+1)}{9x^2-1}$ **g** $\dfrac{3-5x}{(1-x)(2-3x)}$ **h** $\dfrac{x^2-3}{(2+x)(1+x)^2}$

i $\dfrac{5+3x}{(x+2)(x+1)^2}$ **j** $\dfrac{17-5x}{(3+2x)(2-x)^2}$

2 Find the following integrals:

a $\displaystyle\int \frac{2(x^2+3x-1)}{(x+1)(2x-1)}\,\mathrm{d}x$ **b** $\displaystyle\int \frac{x^3+2x^2+2}{x(x+1)}\,\mathrm{d}x$

c $\displaystyle\int \frac{x^2}{x^2-4}\,\mathrm{d}x$ **d** $\displaystyle\int \frac{x^2+x+2}{3-2x-x^2}\,\mathrm{d}x$

e $\displaystyle\int \frac{6+3x-x^2}{x^3+2x^2}\,\mathrm{d}x$

6.5 You can use standard patterns to integrate some expressions.

In Section 6.2 you saw how to integrate $\dfrac{1}{2x+3}$ but you could not apply the technique to integrals of the form $\dfrac{1}{x^2+1}$. However there are families of expressions similar to this that can be integrated easily.

Example 9

Find:

a $\displaystyle\int \frac{2x}{x^2+1}\,\mathrm{d}x$ **b** $\displaystyle\int \frac{\cos x}{3+2\sin x}\,\mathrm{d}x$ **c** $\displaystyle\int 3\cos x\sin^2 x\,\mathrm{d}x$ **d** $\displaystyle\int x(x^2+5)^3\,\mathrm{d}x$

a Let $I = \displaystyle\int \frac{2x}{x^2+1}\,dx$

Consider $y = \ln|x^2+1|$

Then $\dfrac{dy}{dx} = \dfrac{1}{x^2+1} \times 2x$

Remember the $2x$ comes from differentiating x^2+1 using the chain rule.

So $I = \ln|x^2+1| + C$

Since integration is the reverse of differentiation.

b Let $I = \int \frac{\cos x}{3 + 2\sin x}\,dx$

Let $y = \ln|3 + 2\sin x|$

$\frac{dy}{dx} = \frac{1}{3 + 2\sin x} \times 2\cos x$

So $I = \frac{1}{2}\ln|3 + 2\sin x| + C$

Try differentiating $y = \ln|3 + 2\sin x|$.

The $2\cos x$ comes from differentiating $3 + 2\sin x$ with the chain rule.

This is 2 times the required answer so, since integration is the reverse of differentiation you need to divide by 2.

c Let $I = \int 3\cos x \sin^2 x\,dx$

Let $y = \sin^3 x$

$\frac{dy}{dx} = 3\sin^2 x \cos x$

So $I = \sin^3 x + C$

Try differentiating $\sin^3 x$.

The $\cos x$ comes from differentiating $\sin x$ in the chain rule.

d Let $I = \int x(x^2 + 5)^3\,dx$

Let $y = (x^2 + 5)^4$

$\frac{dy}{dx} = 4(x^2 + 5)^3 \times 2x$

$= 8x(x^2 + 5)^3$

So $I = \frac{1}{8}(x^2 + 5)^4 + C$

Try differentiating $(x^2 + 5)^4$.

The $2x$ comes from differentiating $x^2 + 5$.

This is 8 times the required expression so you divide by 8.

You should notice that these examples fall into two types. In **a** and **b** you had $k\frac{f'(x)}{f(x)}$, for some function $f(x)$ and constant k. In **c** and **d** you had $kf'(x)[f(x)]^n$ for some function $f(x)$, constant k and power n.

■ **You should remember the following general patterns:**

- **To integrate expressions of the form $\int k\frac{f'(x)}{f(x)}\,dx$, try $\ln|f(x)|$ and differentiate to check and adjust any constant.**
- **To integrate an expression of the form $\int kf'(x)[f(x)]^n\,dx$, try $[f(x)]^{n+1}$ and differentiate to check and adjust any constant.**

Example 10

Find the following integrals:

a $\int \frac{\text{cosec}^2 x}{(2 + \cot x)^3}\,dx$ **b** $\int 5\tan x \sec^4 x\,dx$

a Let $I = \int \frac{\operatorname{cosec}^2 x}{(2 + \cot x)^3}\,dx$

Notice that $2 + \cot x$ is $f(x)$ and $f'(x) = -\operatorname{cosec}^2 x$, $n = -3$.

Let $y = (2 + \cot x)^{-2}$

$\frac{dy}{dx} = -2(2 + \cot x)^{-3} \times (-\operatorname{cosec}^2 x)$

Use the chain rule.

$= 2(2 + \cot x)^{-3} \operatorname{cosec}^2 x$

This is 2 times the required answer so you need to divide by 2.

So $I = \frac{1}{2}(2 + \cot x)^{-2} + C$

b Let $I = \int 5 \tan x \sec^4 x\,dx$

If $f(x) = \sec x$, then $f'(x)$ is $\sec x \tan x$, so $n = 3$ and $k = 5$.

Let $y = \sec^4 x$

$\frac{dy}{dx} = 4\sec^3 x \times \sec x \tan x$

Use the chain rule.

$= 4\sec^4 x \tan x$

This is $\frac{4}{5}$ times the required answer so you need to divide by $\frac{4}{5}$.

So $I = \frac{5}{4}\sec^4 x + C$

Exercise 6E

1 Integrate the following functions:

a $\frac{x}{x^2 + 4}$ **b** $\frac{e^{2x}}{e^{2x} + 1}$ **c** $\frac{x}{(x^2 + 4)^3}$ **d** $\frac{e^{2x}}{(e^{2x} + 1)^3}$

e $\frac{\cos 2x}{3 + \sin 2x}$ **f** $\frac{\sin 2x}{(3 + \cos 2x)^3}$ **g** xe^{x^2} **h** $\cos 2x(1 + \sin 2x)^4$

i $\sec^2 x \tan^2 x$ **j** $\sec^2 x(1 + \tan^2 x)$

2 Find the following integrals:

a $\int (x + 1)(x^2 + 2x + 3)^4\,dx$ **b** $\int \operatorname{cosec}^2 2x \cot 2x\,dx$

c $\int \sin^5 3x \cos 3x\,dx$ **d** $\int \cos x e^{\sin x}\,dx$

e $\int \frac{e^{2x}}{e^{2x} + 3}\,dx$ **f** $\int x(x^2 + 1)^{\frac{3}{2}}\,dx$

g $\int (2x + 1)\sqrt{x^2 + x + 5}\,dx$ **h** $\int \frac{2x + 1}{\sqrt{x^2 + x + 5}}\,dx$

i $\int \frac{\sin x \cos x}{\sqrt{\cos 2x + 3}}\,dx$ **j** $\int \frac{\sin x \cos x}{\cos 2x + 3}\,dx$

6.6 **Sometimes you can simplify an integral by changing the variable. This process is similar to using the chain rule in differentiation and is called integration by substitution.**

Example 11

Use the substitution $u = 2x + 5$ to find $\int x\sqrt{(2x + 5)}\, dx$.

Let $I = \int x\sqrt{(2x + 5)}\, dx$

You need to replace each 'x' term with a corresponding 'u' term. First replace dx with a term in du.

Let $u = 2x + 5$

So $\frac{du}{dx} = 2$

So d$x = \frac{1}{2}$du.

So 'dx' can be replaced by '$\frac{1}{2}du$'.

$\sqrt{(2x + 5)} = \sqrt{u} = u^{\frac{1}{2}}$

Next rewrite the function in terms of $u = 2x + 5$.

$x = \frac{u - 5}{2}$

Rearrange $u = 2x + 5$ to get $2x = u - 5$ and hence $x = \frac{u - 5}{2}$.

So $I = \int \left(\frac{u - 5}{2}\right) u^{\frac{1}{2}} \times \frac{1}{2}\, du$

Rewrite I in terms of u and simplify.

$= \int \frac{1}{4}(u - 5)u^{\frac{1}{2}}\, du$

$= \int \frac{1}{4}(u^{\frac{3}{2}} - 5u^{\frac{1}{2}})\, du$

$= \frac{1}{4}\frac{u^{\frac{5}{2}}}{\frac{5}{2}} - \frac{5u^{\frac{3}{2}}}{4 \times \frac{3}{2}} + C$

Multiply out the bracket and integrate using rules from your C1 book.

$= \frac{u^{\frac{5}{2}}}{10} - \frac{5u^{\frac{3}{2}}}{6} + C$

Simplify.

So $I = \frac{(2x + 5)^{\frac{5}{2}}}{10} - \frac{5(2x + 5)^{\frac{3}{2}}}{6} + C$

Finally rewrite the answer in terms of x.

Example 12

Use the substitution $u = \sin x + 1$ to find $\int \cos x \sin x\,(1 + \sin x)^3\,dx$.

Let $I = \int \cos x \sin x(1 + \sin x)^3\,dx$

Let $u = \sin x + 1$

$\frac{du}{dx} = \cos x$

So substitute 'cos x dx' with 'du'.

$(\sin x + 1)^3 = u^3$

$\sin x = u - 1$

So $I = \int (u - 1)u^3\,du$

$= \int (u^4 - u^3)\,du$

$= \frac{u^5}{5} - \frac{u^4}{4} + C$

So $I = \frac{(\sin x + 1)^5}{5} - \frac{(\sin x + 1)^4}{4} + C$

First replace the 'dx'.

Notice that this could be split as $du = \cos x\,dx$. The $\cos x$ term can be combined with the dx when substituting.

Use $u = \sin x + 1$ to substitute for the remaining terms, rearranging where required to get $\sin x = u - 1$.

Rewrite I in terms of u.

Multiply out the bracket and integrate in the usual way.

Notice that this example might have been disguised by asking you to use the substitution $u = \sin x + 1$ to find $\int \sin 2x(1 + \sin x)^3\,dx$. You then need to write $\sin 2x$ as $2 \sin x \cos x$ and then proceed as in Example 12. Similar examples might appear in the C4 examination.

In the previous examples the substitution was given. In very simple cases you may be left to choose a substitution of your own.

Example 13

Use integration by substitution to find $\int 6xe^{x^2}\,dx$.

Let $I = \int 6xe^{x^2}\,dx$.

Let $u = x^2$

So $\frac{du}{dx} = 2x$

So replace $2x\,dx$ with du and replace $3e^{x^2}$ with $3e^u$.

So $I = \int 3e^u\,du$

$= 3e^u + C$

So $I = 3e^{x^2} + C$

Since you know how to integrate $\int e^u\,du$, try substituting $u = x^2$.

First aim to replace the dx. You can also use the x and a 2 from the expression. This leaves $3e^{x^2}$.

Write I in terms of u.

Rewrite the answer in terms of x.

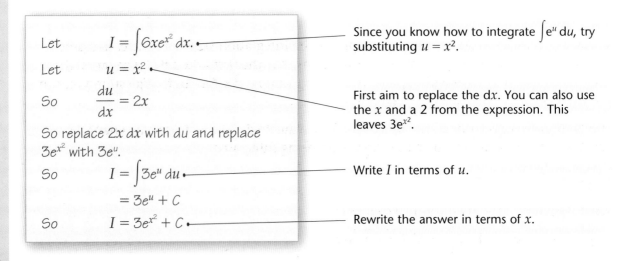

Notice that this integral is one that could have been answered by the methods of Section 6.5. All the integrals in that section can be answered by using substitution but if you can learn how to identify those forms then it is quicker to use the method outlined in that section.

Sometimes implicit differentiation may be needed to help you in the first step of an integration by substitution.

Example 14

Use the substitution $u^2 = 2x + 5$ to find $\int x\sqrt{2x+5}\,dx$. (This solution should be compared with Example 11.)

Let $I = \int x\sqrt{2x+5}\,dx$ — First aim to replace the dx.

$u^2 = 2x + 5$

$2u\dfrac{du}{dx} = 2$ — Using implicit differentiation, cancel 2 and rearrange to get d$x = u$du.

So replace dx with udu.

$\sqrt{2x+5} = u$ — Substitute the remaining expressions. You will need to make x the subject of $u^2 = 2x + 5$.

and $x = \dfrac{u^2 - 5}{2}$

So $I = \int\left(\dfrac{u^2-5}{2}\right)u \times udu$

$= \int\left(\dfrac{u^4}{2} - \dfrac{5u^2}{2}\right)du$ — Multiply out the brackets and integrate.

$= \dfrac{u^5}{10} - \dfrac{5u^3}{6} + C$

So $I = \dfrac{(2x+5)^{\frac{5}{2}}}{10} - \dfrac{5(2x+5)^{\frac{3}{2}}}{6} + C$ — Rewrite answer in terms of x.

If you compare this solution with Example 11 the integration step is a little easier (since you are dealing with integer powers not fractions) but the first and last steps you might consider a little more difficult. Unless the substitution is specified in the question you can choose which sort of substitution you wish to use.

Integration by substitution can also be used to evaluate definite integrals by changing the limits of the integral as well as the expression being integrated.

Example 15

Use integration by substitution to evaluate:

a $\int_0^2 x(x+1)^3\,dx$ **b** $\int_0^{\frac{\pi}{2}} \cos x\sqrt{1+\sin x}\,dx$

a Let $I = \int_0^2 x(x+1)^3\,dx$

Let $u = x + 1$

$\frac{du}{dx} = 1$

so replace dx with du and replace $(x+1)^3$ with u^3, and x with $u - 1$.

Replace each term in x with a term in u in the usual way.

x	u
2	3
0	1

Change the limits. When $x = 2$, $u = 2 + 1 = 3$ and when $x = 0$, $u = 1$.

So $I = \int_1^3 (u-1)u^3\,du$

Note that the new u limits replace their corresponding x limits.

$= \int_1^3 (u^4 - u^3)\,du$

$= \left[\frac{u^5}{5} - \frac{u^4}{4}\right]_1^3$

Multiply out and integrate. Remember there is no need for a $+C$.

$= \left(\frac{243}{5} - \frac{81}{4}\right) - \left(\frac{1}{5} - \frac{1}{4}\right)$

The integral can now be evaluated using the limits for u without having to change back into x.

$= 48.4 - 20 = 28.4$

b $\int_0^{\frac{\pi}{2}} \cos x\sqrt{1 + \sin x}\,dx$

Use $u = 1 + \sin x$.

$u = 1 + \sin x \Rightarrow \frac{du}{dx} = \cos x$, so replace $\cos x\,dx$ with du and replace $\sqrt{(1 + \sin x)}$ with $u^{\frac{1}{2}}$.

Remember that limits for integrals involving trigonometric functions will always be in radians.

x	u
$\frac{\pi}{2}$	2
0	1

$x = \frac{\pi}{2}$, means $u = 1 + 1 = 2$ and $x = 0$, means $u = 1 + 0 = 1$.

So $I = \int_1^2 u^{\frac{1}{2}}\,du$

Rewrite the integral in terms of u.

$= \left[\frac{2}{3}u^{\frac{3}{2}}\right]_1^2$

$= \left(\frac{2}{3}2^{\frac{3}{2}}\right) - \left(\frac{2}{3}\right)$

Remember that $2^{\frac{3}{2}} = \sqrt{8} = 2\sqrt{2}$.

So $I = \frac{2}{3}(2\sqrt{2} - 1)$

Exercise 6F

1 Use the given substitution to find the following integrals:

a $\int x\sqrt{1+x}\,dx;\ u = 1 + x$

b $\int \frac{x}{\sqrt{1+x}}\,dx;\ u = 1 + x$

c $\int \frac{1+\sin x}{\cos x}\,dx;\ u = \sin x$

d $\int x(3+2x)^5\,dx;\ u = 3 + 2x$

e $\int \sin^3 x\,dx;\ u = \cos x$

2 Use the given substitution to find the following integrals:

a $\int x\sqrt{2+x}\,dx;\ u^2 = 2 + x$

b $\int \frac{2}{\sqrt{x}(x-4)}\,dx;\ u = \sqrt{x}$

c $\int \sec^2 x \tan x\sqrt{1+\tan x}\,dx;\ u^2 = 1 + \tan x$

d $\int \frac{\sqrt{x^2+4}}{x}\,dx;\ u^2 = x^2 + 4$

e $\int \sec^4 x\,dx;\ u = \tan x$

3 Evaluate the following:

a $\int_0^5 x\sqrt{x+4}\,dx$

b $\int_0^{\frac{\pi}{3}} \sec x \tan x\sqrt{\sec x + 2}\,dx$

c $\int_2^5 \frac{1}{1+\sqrt{x-1}}\,dx$; use $u^2 = x - 1$

d $\int_0^{\frac{\pi}{2}} \frac{\sin 2\theta}{1+\cos\theta}\,d\theta$; let $u = 1 + \cos\theta$

e $\int_0^1 x(2+x)^3\,dx$

f $\int_1^4 \frac{1}{\sqrt{x}(4x-1)}\,dx$; let $u = \sqrt{x}$

6.7 You can use integration by parts to integrate some expressions.

In the C3 book you met the product rule for differentiation:

$$\frac{d}{dx}(uv) = v\frac{du}{dx} + u\frac{dv}{dx}$$

Rearranging gives

$$u\frac{dv}{dx} = \frac{d}{dx}(uv) - v\frac{du}{dx}$$

Now integrating each term with respect to x gives

$$\int u\frac{dv}{dx}\,dx = \int \frac{d}{dx}(uv)\,dx - \int v\frac{du}{dx}\,dx$$

Now, since differentiating a function and then integrating it leaves the function unchanged, you can simplify $\int \frac{d}{dx}(uv)\,dx$ to uv, and this gives the **integration by parts** formula:

■ $$\int u\frac{dv}{dx}\,dx = uv - \int v\frac{du}{dx}\,dx$$

This formula enables you to exchange an integral which is complicated $\left(\int u\frac{dv}{dx}dx\right)$ for a simpler one $\left(\int v\frac{du}{dx}dx\right)$. You will not be expected to produce this proof in the C4 examination.

Example 16

Find $\int x\cos x\,dx$.

Let $I = \int x\cos x\,dx$

$$u = x \quad \Rightarrow \frac{du}{dx} = 1$$

$$v = \sin x \Leftarrow \frac{dv}{dx} = \cos x$$

Using the integration by parts formula:

$$I = x\sin x - \int \sin x \times 1\,dx$$

$$= x\sin x + \cos x + C$$

Let $u = x$ and $\frac{dv}{dx} = \cos x$.

Complete the table for u, v, $\frac{du}{dx}$ and $\frac{dv}{dx}$.

Take care to differentiate u but integrate $\frac{dv}{dx}$.

Notice that $\int v\frac{du}{dx}dx$ is a simpler integral than $\int u\frac{dv}{dx}dx$.

In general you will *usually* let u = any terms of the form x^n, but there is one exception to this and that is when there is a $\ln x$ term. In this case you should let u = the $\ln x$ term.

Example 17

Find $\int x^2\ln x\,dx$.

Let $I = \int x^2\ln x\,dx$

$$u = \ln x \Rightarrow \frac{du}{dx} = \frac{1}{x}$$

$$v = \frac{x^3}{3} \Leftarrow \frac{dv}{dx} = x^2$$

$$I = \frac{x^3}{3}\ln x - \int \frac{x^3}{3} \times \frac{1}{x}dx$$

$$= \frac{x^3}{3}\ln x - \int \frac{x^2}{3}dx$$

$$= \frac{x^3}{3}\ln x - \frac{x^3}{9} + C$$

Since there is a $\ln x$ term, let $u = \ln x$ and $\frac{dv}{dx} = x^2$.

Complete the table for u, v, $\frac{du}{dx}$ and $\frac{dv}{dx}$.

Take care to differentiate u but integrate $\frac{dv}{dx}$.

Apply the integration by parts formula.

Simplify the $v\frac{du}{dx}$ term.

Sometimes you may have to use integration by parts twice.

Example 18

Find $\int x^2 e^x \, dx$.

Let $I = \int x^2 e^x \, dx$

$u = x^2 \Rightarrow \frac{du}{dx} = 2x$

$v = e^x \Leftarrow \frac{dv}{dx} = e^x$

So $I = x^2 e^x - \int 2x e^x \, dx$

$u = 2x \Rightarrow \frac{du}{dx} = 2$

$v = e^x \Leftarrow \frac{dv}{dx} = e^x$

So $I = x^2 e^x - \left[2x e^x - \int 2e^x \, dx\right]$

$= x^2 e^x - 2xe^x + \int 2e^x \, dx$

$= x^2 e^x - 2xe^x + 2e^x + C$

There is no $\ln x$ term so let $u = x^2$ and $\frac{dv}{dx} = e^x$.

Complete the table for u, v, $\frac{du}{dx}$ and $\frac{dv}{dx}$.

Take care to differentiate u but integrate $\frac{dv}{dx}$.

Apply the integration by parts formula.

Notice that this integral is simpler than I but still not one you can write down. It has a similar structure to I and so you can use integration by parts again with $u = 2x$ and $\frac{dv}{dx} = e^x$.

Apply the integration by parts formula a second time.

Integration by parts involves integrating in two separate stages (first the uv term then the $\int v \frac{du}{dx} \, dx$). Any limits can be applied separately to each part.

Example 19

Evaluate $\int_1^2 \ln x \, dx$, leaving your answer in terms of natural logarithms.

Let $I = \int_1^2 \ln x \, dx = \int_1^2 \ln x \times 1 \, dx$

$u = \ln x \Rightarrow \frac{du}{dx} = \frac{1}{x}$

$v = x \Leftarrow \frac{dv}{dx} = 1$

$I = [x \ln x]_1^2 - \int_1^2 x \times \frac{1}{x} \, dx$

$= (2 \ln 2) - (1 \ln 1) - \int_1^2 1 \, dx$

$= 2 \ln 2 - [x]_1^2$

$= 2 \ln 2 - 2 - 1$

$= 2 \ln 2 - 1$

Write the expression to be integrated as $\ln x \times 1$, then $u = \ln x$ and $\frac{dv}{dx} = 1$.

Complete the usual table.

Apply the limits to the uv term and to $\int v \frac{du}{dx} \, dx$.

Evaluate the limits on uv and remember $\ln 1 = 0$.

- **The following integrals should be given in your formulae booklet and they can easily be verified by differentiation. Some of them are used in the next exercise.**
  - $\int \tan x \, dx = \ln|\sec x| + C$
  - $\int \sec x \, dx = \ln|\sec x + \tan x| + C$
  - $\int \cot x \, dx = \ln|\sin x| + C$
  - $\int \text{cosec}\, x \, dx = -\ln|\text{cosec}\, x + \cot x| + C$

Exercise 6G

1 Find the following integrals:

a $\int x \sin x \, dx$ **b** $\int x e^x \, dx$ **c** $\int x \sec^2 x \, dx$

d $\int x \sec x \tan x \, dx$ **e** $\int \frac{x}{\sin^2 x} \, dx$

2 Find the following integrals:

a $\int x^2 \ln x \, dx$ **b** $\int 3 \ln x \, dx$ **c** $\int \frac{\ln x}{x^3} \, dx$

d $\int (\ln x)^2 \, dx$ **e** $\int (x^2 + 1) \ln x \, dx$

3 Find the following integrals:

a $\int x^2 e^{-x} \, dx$ **b** $\int x^2 \cos x \, dx$ **c** $\int 12x^2(3 + 2x)^5 \, dx$

d $\int 2x^2 \sin 2x \, dx$ **e** $\int x^2 2 \sec^2 x \tan x \, dx$

4 Evaluate the following:

a $\int_0^{\ln 2} x e^{2x} \, dx$ **b** $\int_0^{\frac{\pi}{2}} x \sin x \, dx$ **c** $\int_0^{\frac{\pi}{2}} x \cos x \, dx$ **d** $\int_1^2 \frac{\ln x}{x^2} \, dx$

e $\int_0^1 4x(1 + x)^3 \, dx$ **f** $\int_0^{\pi} x \cos(\tfrac{1}{4}x) \, dx$ **g** $\int_0^{\frac{\pi}{3}} \sin x \ln(\sec x) \, dx$

6.8 You can use numerical integration.

In the C2 book you met the **trapezium rule** for finding an approximate value for a definite integral. In C4 you may be asked to use the trapezium rule for integrals involving some of the new functions met in C3 and C4.

- **Remember: the trapezium rule is**

$$\int_a^b y \, dx \approx \tfrac{1}{2} h[y_0 + 2(y_1 + y_2 + \ldots + y_{n-1}) + y_n]$$

where $h = \dfrac{b - a}{n}$ **and** $y_i = f(a + ih)$

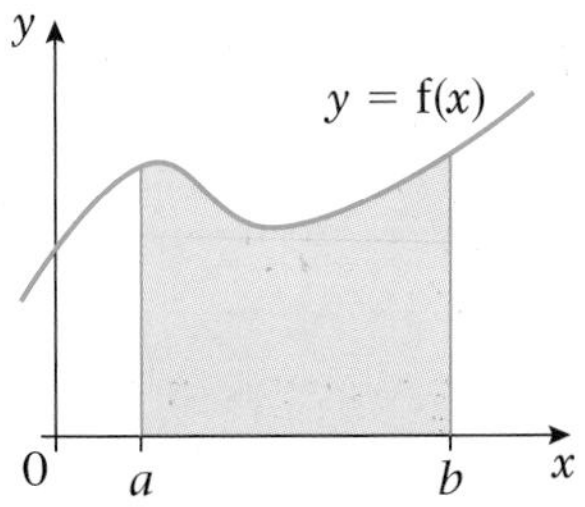

Example 20

For the integral $I = \int_0^{\frac{\pi}{3}} \sec x \, dx$:

a Find the exact value of I.

b Use the trapezium rule with two strips to estimate I.

c Use the trapezium rule with four strips to find a second estimate for I.

d Find the percentage error in using these two estimates for I.

a $I = \int_0^{\frac{\pi}{3}} \sec x \, dx$:

$= \left[\ln|\sec x + \tan x| \right]_0^{\frac{\pi}{3}}$

$= (\ln|2 + \sqrt{3}|) - (\ln|1 + 0|)$

$= \ln(2 + \sqrt{3})$

Use the formulae booklet or see Section 6.7.

$\sec\frac{\pi}{3} = \frac{1}{\cos\frac{\pi}{3}} = \frac{1}{0.5} = 2$ and $\tan\frac{\pi}{3} = \sqrt{3}$.

b

x	0	$\frac{\pi}{6}$	$\frac{\pi}{3}$
y	1	1.155	2

Complete the table to find the values of y.

$I \approx \frac{1}{2}\frac{\pi}{6}[1 + 2 \times 1.155 + 2]$

$= \frac{\pi}{12} \times 5.31 = 1.390\ldots$

$= 1.39$ (3 s.f.)

c

x	0	$\frac{\pi}{12}$	$\frac{\pi}{6}$	$\frac{\pi}{4}$	$\frac{\pi}{3}$
y	1	1.035	1.155	1.414	2

Complete the table to find the values of y.

$I \approx \frac{1}{2}\frac{\pi}{12}[1 + 2(1.035 + 1.155 + 1.414) + 2]$

$= \frac{\pi}{24}[10.208]$

$= 1.336\,224\,075\ldots = 1.34$ (3 s.f.)

d

Percentage error in using **b** is

$$\frac{(1.390\ldots - \ln|2+\sqrt{3}|)}{\ln|2+\sqrt{3}|} \times 100 = 5.6\%$$

Percentage error in using **c** is

$$\frac{(1.336\ldots - \ln|2+\sqrt{3}|)}{\ln|2+\sqrt{3}|} \times 100 = 1.5\%$$

So **c** is more accurate.

Exercise 6H

1 Use the trapezium rule with n strips to estimate the following:

a $\int_0^3 \ln(1+x^2)\,dx;\ n = 6$

b $\int_0^{\frac{\pi}{3}} \sqrt{(1+\tan x)}\,dx;\ n = 4$

c $\int_0^2 \frac{1}{\sqrt{(e^x+1)}}\,dx;\ n = 4$

d $\int_{-1}^{1} \text{cosec}^2(x^2+1)\,dx;\ n = 4$

e $\int_{0.1}^{1.1} \sqrt{\cot x}\,dx;\ n = 5$

2 **a** Find the exact value of $I = \int_1^4 x \ln x\,dx$.

b Find approximate values for I using the trapezium rule with

i 3 strips **ii** 6 strips

c Compare the percentage error for these two approximations.

3 **a** Find an approximate value for $I = \int_0^1 e^x \tan x\,dx$ using

i 2 **ii** 4 **iii** 8 strips.

b Suggest a possible value for I.

4 **a** Find the exact value of $I = \int_0^2 x\sqrt{(2-x)}\,dx$.

b Find an approximate value for I using the trapezium rule with

i 4 and **ii** 6 strips.

c Compare the percentage error for these two approximations.

6.9 You can use integration to find areas and volumes.

In the C2 book you saw how to find the area of a region R between a curve and the x-axis:

■ **Area of region between $y = \mathrm{f}(x)$, the x-axis and $x = a$ and $x = b$ is given by:**

$$\text{Area} = \int_a^b y\,\mathrm{d}x$$

This area can be thought of as a limit of a sum of approximate rectangular strips of width δx and length y.

Hint: Since the strip is roughly rectangular the area is approximately $y\delta x$.

Thus the area is the limit of $\sum y\delta x$ as $\delta x \to 0$. The integration symbol $\int$ is an elongated 'S' to represent this idea of a sum.

If each strip is now revolved through 2π radians (or 360 degrees) about the x-axis, it will form a shape that is approximately cylindrical. The volume of each cylinder will be $\pi y^2 \delta x$ since the radius is y and the height is δx.

The limit of the sum $\sum \pi y^2 \delta x$, as $\delta x \to 0$, is given by $\pi \int y^2\,\mathrm{d}x$ and this formula can be used to find the volume of the solid formed when the region R is rotated through 2π radians about the x-axis.

■ **Volume of revolution formed when $y = \mathrm{f}(x)$ is rotated about the x-axis between $x = a$ and $x = b$ is given by:**

$$\text{Volume} = \pi \int y^2\,\mathrm{d}x$$

Example 21

The region R is bounded by the curve with equation $y = \sin 2x$, the x-axis and the lines $x = 0$ and $x = \frac{\pi}{2}$.

a Find the area of R.

b Find the volume of the solid formed when the region R is rotated through 2π radians about the x-axis.

a $\text{Area} = \int_0^{\frac{\pi}{2}} \sin 2x\,\mathrm{d}x$

$$= \left[-\tfrac{1}{2}\cos 2x\right]_0^{\frac{\pi}{2}}$$

$$= \left(-\tfrac{1}{2}(-1)\right) - \left(-\tfrac{1}{2}\right)$$

$$= 1$$

b Volume $= \pi\int_0^{\frac{\pi}{2}} \sin^2 2x\,dx$

$= \pi\int_0^{\frac{\pi}{2}} \frac{1}{2}(1 - \cos 4x)\,dx$

$= \pi\left[\frac{1}{2}x - \frac{1}{8}\sin 4x\right]_0^{\frac{\pi}{2}}$

$= \left(\frac{\pi^2}{4} - 0\right) - (0)$

$= \frac{\pi^2}{4}$

Use $\cos 2A = 1 - 2\sin^2 A$.

Rearrange to give $\sin^2 A = \ldots$

Note that $2 \times 2x$ gives $4x$ in the cos term.

Multiply out and integrate.

Sometimes the equation of the curve may be given in terms of parameters. You can integrate in terms of the parameter by changing the variable in a manner similar to that described in Section 6.6.

Example 22

The curve C has parametric equations

$$x = t(1 + t)$$
$$y = \frac{1}{1 + t}$$

where t is the parameter and $t \geqslant 0$.
The region R is bounded by C, the x-axis and the lines $x = 0$ and $x = 2$.

a Find the exact area of R.

b Find the exact volume of the solid formed when R is rotated through 2π radians about the x-axis.

a Area $= \int_0^2 y\,dx$

By the chain rule $\int y\,dx = \int y\frac{dx}{dt}\,dt$

$x = t(1 + t) \Rightarrow \frac{dx}{dt} = 1 + 2t$

$x = 0$ so $t(1 + t) = 0$ so $t = 0$ or -1,
but since $t \geqslant 0$, $t = 0$

$x = 2$ so $t^2 + t - 2 = 0$

so $(t + 2)(t - 1) = 0$, so $t = 1$ or -2,
but since $t \geqslant 0$, $t = 1$

You need to change the integral into one in terms of t.

Write $x = t + t^2$ and then differentiate.

Change the limits.

So Area $= \int y \frac{dx}{dt}\, dt$

$= \int_0^1 \frac{1}{(1+t)}(1+2t)\, dt$ — Divide $(1 + t)$ into the numerator.

$= \int_0^1 \left(2 - \frac{1}{1+t}\right) dt$ — Simplify using algebraic division.

$= \Big[2t - \ln|1+t|\Big]_0^1$

$= (2 - \ln 2) - (0 - \ln 1)$

$= 2 - \ln 2$

b Volume $= \pi \int y^2\, dx = \pi \int y^2 \frac{dx}{dt}\, dt$ — By the chain rule as above.

Limits will be the same as for the area.

Volume $= \pi \int_0^1 \frac{1}{(1+t)^2}(1+2t)\, dt$

$\frac{1+2t}{(1+t)^2} \equiv \frac{A}{(1+t)^2} + \frac{B}{(1+t)}$ — Use partial fractions.

$1 + 2t = A + B(1 + t)$

$B = 2$

$A = -1$

Substitute values of t or compare coefficients.

So Volume $= \int_0^1 \left(\frac{2}{1+t} - \frac{1}{(1+t)^2}\right) dt$

$= \pi\left[2\ln|1+t| + \frac{1}{(1+t)}\right]_0^1$

$= \pi[(2\ln 2 + \tfrac{1}{2}) - (0 + 1)]$

$= \pi(2\ln 2 - \tfrac{1}{2})$

Exercise 6I

1 The region R is bounded by the curve with equation $y = f(x)$, the x-axis and the lines $x = a$ and $x = b$. In each of the following cases find the exact value of:

i the area of R,

ii the volume of the solid of revolution formed by rotating R through 2π radians about the x-axis.

a $f(x) = \frac{2}{1+x}$; $a = 0$, $b = 1$

b $f(x) = \sec x$; $a = 0$, $b = \frac{\pi}{3}$

c $f(x) = \ln x$; $a = 1$, $b = 2$

d $f(x) = \sec x \tan x$; $a = 0$, $b = \frac{\pi}{4}$

e $f(x) = x\sqrt{4 - x^2}$; $a = 0$, $b = 2$

2 Find the exact area between the curve $y = \mathrm{f}(x)$, the x-axis and the lines $x = a$ and $x = b$ where:

a $\mathrm{f}(x) = \dfrac{4x + 3}{(x + 2)(2x - 1)}$; $a = 1$, $b = 2$

b $\mathrm{f}(x) = \dfrac{x}{(x + 1)^2}$; $a = 0$, $b = 2$

c $\mathrm{f}(x) = x \sin x$; $a = 0$, $b = \dfrac{\pi}{2}$

d $\mathrm{f}(x) = \cos x \sqrt{2 \sin x + 1}$; $a = 0$, $b = \dfrac{\pi}{6}$

e $\mathrm{f}(x) = x\mathrm{e}^{-x}$; $a = 0$, $b = \ln 2$

3 The region R is bounded by the curve C, the x-axis and the lines $x = -8$ and $x = +8$. The parametric equations for C are $x = t^3$ and $y = t^2$. Find:

a the area of R,

b the volume of the solid of revolution formed when R is rotated through 2π radians about the x-axis.

4 The curve C has parametric equations $x = \sin t$, $y = \sin 2t$, $0 \leqslant t \leqslant \dfrac{\pi}{2}$.

a Find the area of the region bounded by C and the x-axis.

If this region is revolved through 2π radians about the x-axis,

b find the volume of the solid formed.

6.10 You can use integration to solve differential equations.

In Chapter 4 you met differential equations. In this section you will learn how to solve simple first order differential equations by the process called **separation of variables**.

- **When $\dfrac{\mathrm{d}y}{\mathrm{d}x} = \mathrm{f}(x)g(y)$ you can write**

$$\int \frac{1}{g(y)}\,\mathrm{d}y = \int \mathrm{f}(x)\,\mathrm{d}x$$

This is called separating the variables.

Example 23

Find a general solution to the differential equation $(1 + x^2)\dfrac{\mathrm{d}y}{\mathrm{d}x} = x \tan y$.

$$\frac{\mathrm{d}y}{\mathrm{d}x} = \frac{x}{1 + x^2}\tan y$$

Write the equation in the form $\dfrac{\mathrm{d}y}{\mathrm{d}x} = \mathrm{f}(x)\mathrm{g}(y)$

$$\int \frac{1}{\tan y}\,\mathrm{d}y = \int \frac{x}{1 + x^2}\,\mathrm{d}x$$

Now **separate the variables** so that $\dfrac{1}{\mathrm{g}(y)}\,\mathrm{d}y = \mathrm{f}(x)\,\mathrm{d}x$.

$$\int \cot y\,\mathrm{d}y = \int \frac{x}{1 + x^2}\,\mathrm{d}x$$

Use $\cot y = \dfrac{1}{\tan y}$.

$$\ln|\sin y| = \tfrac{1}{2}\ln|1 + x^2| + C$$

Integrate, remembering that the integral of $\cot y$ is in the formulae book (or see page 102).

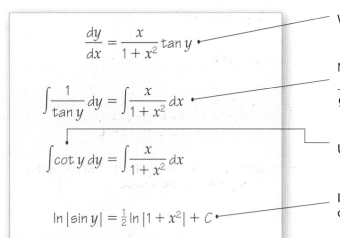

or $\ln|\sin y| = \frac{1}{2}\ln|1 + x^2| + \ln k$ — Don't forget the $+C$ which could be written as $\ln k$.

$\ln|\sin y| = \ln|k\sqrt{1 + x^2}|$ — Combining logs.

so $\sin y = k\sqrt{1 + x^2}$ — Finally remove the ln. Sometimes you might be asked to give your answer in the form $y = f(x)$. This question did not specify that so it is acceptable to give the answer in this form.

Sometimes **boundary conditions** are given in a question which enable you to find a **particular** solution to the differential equation. In this case you first find the general solution and then apply the boundary conditions to find the value of the constant of integration.

Example 24

Find the particular solution of the differential equation

$$\frac{dy}{dx} = \frac{-3(y-2)}{(2x+1)(x+2)}$$

given that $x = 1$ when $y = 4$. Leave your answer in the form $y = f(x)$.

$$\int \frac{1}{y-2}\,dy = \int \frac{-3}{(2x+1)(x+2)}\,dx$$

First separate the variables.

$$\frac{-3}{(2x+1)(x+2)} \equiv \frac{A}{(2x+1)} + \frac{B}{(x+2)}$$

Use partial fractions for this integral.

$$-3 = A(x+2) + B(2x+1)$$

$x = -2$ gives $-3 = -3B$ so $B = 1$

$x = -0.5$ gives $-3 = \frac{3}{2}A$ so $A = -2$

So

$$\int \frac{1}{y-2}\,dy = \int\left(\frac{1}{x+2} - \frac{2}{2x+1}\right)dx$$

Rewrite the integral using the partial fractions.

$$\ln|y-2| = \ln|x+2| - \ln|2x+1| + \ln k$$

Integrate and use $+\ln k$ instead of $+C$.

$$\ln|y-2| = \ln\left|\frac{k(x+2)}{(2x+1)}\right|$$

Combine ln terms.

$$y - 2 = k\left(\frac{x+2}{2x+1}\right)$$

Remove ln.

$$4 - 2 = k\left(\frac{1+2}{2+1}\right) \Rightarrow k = 2$$

Use the condition $x = 1$ when $y = 4$ by substituting these values in the general solution here, and solve to find k.

So
$$y = 2 + 2\left(\frac{x+2}{2x+1}\right)$$
$$= 3 + \frac{3}{2x+1}$$

Substitute $k = 2$ and write the answer in the form $y = f(x)$ as requested.

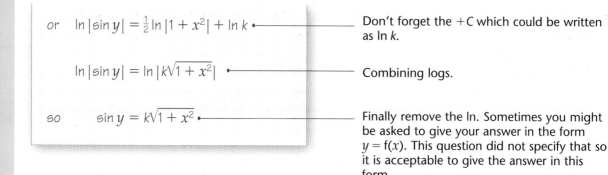

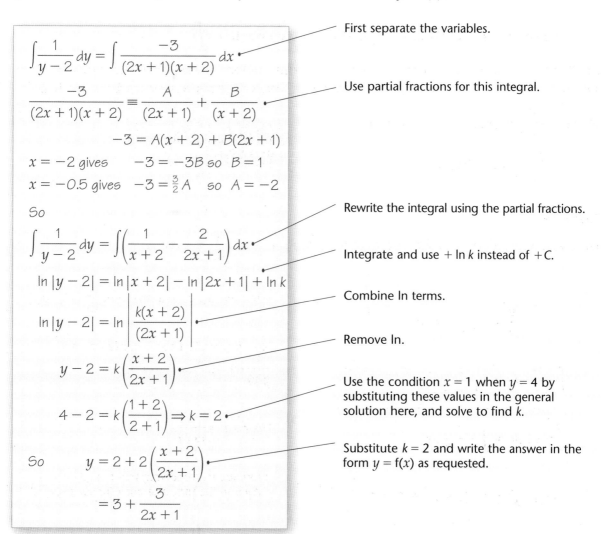

Exercise 6J

1 Find general solutions of the following differential equations. Leave your answer in the form $y = \mathrm{f}(x)$.

a $\dfrac{dy}{dx} = (1 + y)(1 - 2x)$

b $\dfrac{dy}{dx} = y \tan x$

c $\cos^2 x \dfrac{dy}{dx} = y^2 \sin^2 x$

d $\dfrac{dy}{dx} = 2e^{x-y}$

e $x^2 \dfrac{dy}{dx} = y + xy$

2 Find a general solution of the following differential equations. (You do not need to write the answers in the form $y = \mathrm{f}(x)$.)

a $\dfrac{dy}{dx} = \tan y \tan x$

b $\sin y \cos x \dfrac{dy}{dx} = \dfrac{x \cos y}{\cos x}$

c $(1 + x^2)\dfrac{dy}{dx} = x(1 - y^2)$

d $\cos y \sin 2x \dfrac{dy}{dx} = \cot x \operatorname{cosec} y$

e $e^{x+y}\dfrac{dy}{dx} = x(2 + e^y)$

3 Find general solutions of the following differential equations:

a $\dfrac{dy}{dx} = ye^x$

b $\dfrac{dy}{dx} = xe^y$

c $\dfrac{dy}{dx} = y \cos x$

d $\dfrac{dy}{dx} = x \cos y$

e $\dfrac{dy}{dx} = (1 + \cos 2x) \cos y$

f $\dfrac{dy}{dx} = (1 + \cos 2y) \cos x$

4 Find particular solutions of the following differential equations using the given boundary conditions.

a $\dfrac{dy}{dx} = \sin x \cos^2 x;\ y = 0,\ x = \dfrac{\pi}{3}$

b $\dfrac{dy}{dx} = \sec^2 x \sec^2 y;\ y = 0,\ x = \dfrac{\pi}{4}$

c $\dfrac{dy}{dx} = 2\cos^2 y \cos^2 x;\ y = \dfrac{\pi}{4},\ x = 0$

d $(1 - x^2)\dfrac{dy}{dx} = xy + y;\ x = 0.5,\ y = 6$

e $2(1 + x)\dfrac{dy}{dx} = 1 - y^2;\ x = 5,\ y = \frac{1}{2}$

6.11 Sometimes the differential equation will arise out of a context and the solution may need interpreting in terms of that context.

Example 25

The rate of increase of a population P of microorganisms at time t is given by $\frac{dP}{dt} = kP$, where k is a positive constant. Given that at $t = 0$ the population was of size 8, and at $t = 1$ the population is 56, find the size of the population at time $t = 2$.

$$\frac{dP}{dt} = kP$$

Separate the variables.

$$\int \frac{1}{P}\,dP = \int k\,dt$$

Integrate and include the $+C$.

$$\ln|P| = kt + C$$

Notice that there are 2 unknown letters, k and C. Use $t = 0$ and $P = 8$ to find C.

$t = 0$, $P = 8$ gives $\ln 8 = C$

So $\ln|P| = kt + \ln 8$

Or $\ln\left|\frac{P}{8}\right| = kt$

Use properties of logs to simplify.

$t = 1$, $P = 56$ gives $\ln 7 = k$

Now use $t = 1$ and $P = 56$ to find k.

So $\ln\left|\frac{P}{8}\right| = t\ln 7$

Now let $t = 2$.

$t = 2$ gives

$$\ln\left|\frac{P}{8}\right| = 2\ln 7$$

but $2\ln 7 = \ln 49 \Rightarrow \frac{P}{8} = 49$

So $P = 8 \times 49 = 392$

Many of the examples in the next exercise are related to differential equations met in Section 4.5.

Exercise 6K

1 The size of a certain population at time t is given by P. The rate of increase of P is given by $\frac{dP}{dt} = 2P$. Given that at time $t = 0$, the population was 3, find the population at time $t = 2$.

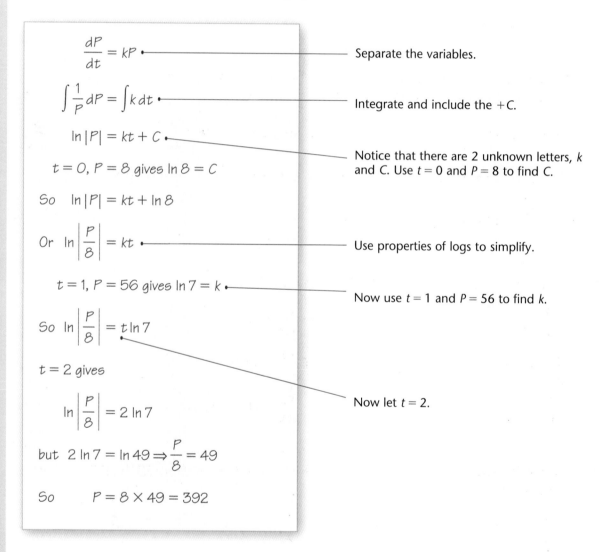

2 The number of particles at time t of a certain radioactive substance is N. The substance is decaying in such a way that $\frac{dN}{dt} = -\frac{N}{3}$.

Given that at time $t = 0$ the number of particles is N_0, find the time when the number of particles remaining is $\frac{1}{2}N_0$.

3 The mass M at time t of the leaves of a certain plant varies according to the differential equation $\frac{dM}{dt} = M - M^2$.

a Given that at time $t = 0$, $M = 0.5$, find an expression for M in terms of t.

b Find a value for M when $t = \ln 2$.

c Explain what happens to the value of M as t increases.

4 The volume of liquid $V\,\text{cm}^3$ at time t seconds satisfies

$$-15\frac{dV}{dt} = 2V - 450.$$

Given that initially the volume is $300\,\text{cm}^3$, find to the nearest cm^3 the volume after 15 seconds.

5 The thickness of ice x mm on a pond is increasing and $\frac{dx}{dt} = \frac{1}{20x^2}$, where t is measured in hours. Find how long it takes the thickness of ice to increase from 1 mm to 2 mm.

6 The depth h metres of fluid in a tank at time t minutes satisfies $\frac{dh}{dt} = -k\sqrt{h}$, where k is a positive constant. Find, in terms of k, how long it takes the depth to decrease from 9 m to 4 m.

7 The rate of increase of the radius r kilometres of an oil slick is given by $\frac{dr}{dt} = \frac{k}{r^2}$, where k is a positive constant. When the slick was first observed the radius was 3 km. Two days later it was 5 km. Find, to the nearest day when the radius will be 6.

Mixed exercise 6L

1 It is given that $y = x^{\frac{3}{2}} + \frac{48}{x}$, $x > 0$.

a Find the value of x and the value of y when $\frac{dy}{dx} = 0$.

b Show that the value of y which you found is a minimum.

The finite region R is bounded by the curve with equation $y = x^{\frac{3}{2}} + \frac{48}{x}$, the lines $x = 1$, $x = 4$ and the x-axis.

c Find, by integration, the area of R giving your answer in the form $p + q\ln r$, where the numbers p, q and r are to be found. **E**

2 The curve C has two arcs, as shown, and the equations

$$x = 3t^2,\ y = 2t^3,$$

where t is a parameter.

a Find an equation of the tangent to C at the point P where $t = 2$.

The tangent meets the curve again at the point Q.

b Show that the coordinates of Q are $(3, -2)$.

The shaded region R is bounded by the arcs OP and OQ of the curve C, and the line PQ, as shown.

c Find the area of R.

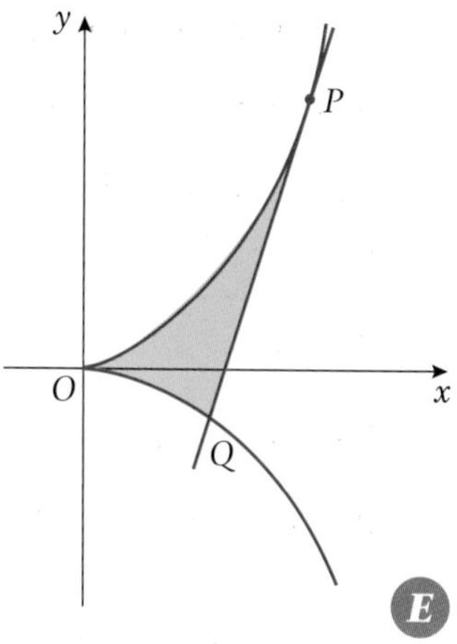

E

3 **a** Show that $(1 + \sin 2x)^2 \equiv \frac{1}{2}(3 + 4\sin 2x - \cos 4x)$.

b The finite region bounded by the curve with equation $y = 1 + \sin 2x$, the x-axis, the y-axis and the line with equation $x = \dfrac{\pi}{2}$ is rotated through 2π about the x-axis.

Using calculus, calculate the volume of the solid generated, giving your answer in terms of π.

E

4

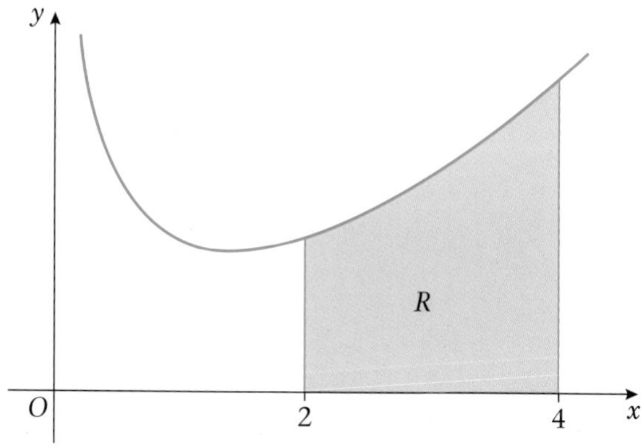

This graph shows part of the curve with equation $y = \mathrm{f}(x)$ where

$$\mathrm{f}(x) \equiv \mathrm{e}^{0.5x} + \frac{1}{x},\ x > 0.$$

The curve has a stationary point at $x = \alpha$.

a Find $\mathrm{f}'(x)$.

b Hence calculate $\mathrm{f}'(1.05)$ and $\mathrm{f}'(1.10)$ and deduce that $1.05 < \alpha < 1.10$.

c Find $\int \mathrm{f}(x)\,\mathrm{d}x$.

The shaded region R is bounded by the curve, the x-axis and the lines $x = 2$ and $x = 4$.

d Find, to 2 decimal places, the area of R.

E

5 **a** Find $\int x\mathrm{e}^{-x}\,\mathrm{d}x$.

b Given that $y = \dfrac{\pi}{4}$ at $x = 0$, solve the differential equation

$$\mathrm{e}^x \frac{\mathrm{d}y}{\mathrm{d}x} = \frac{x}{\sin 2y}.$$

E

6 The diagram shows the finite shaded region bounded by the curve with equation $y = x^2 + 3$, the lines $x = 1$, $x = 0$ and the x-axis. This region is rotated through 360° about the x-axis.

Find the volume generated.

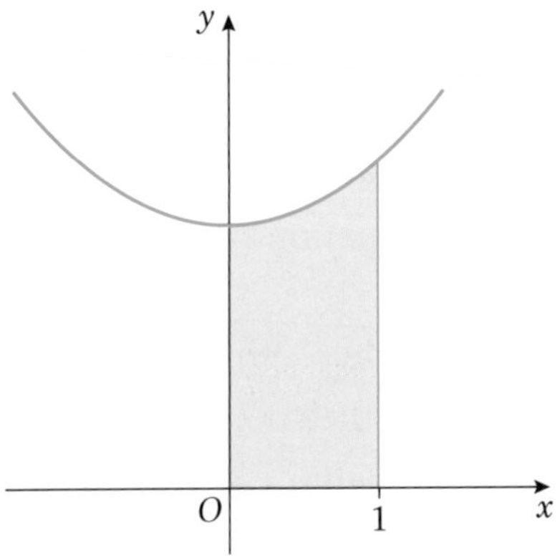

7 **a** Find $\int \frac{1}{x(x+1)}\,dx$

b Using the substitution $u = e^x$ and the answer to **a**, or otherwise, find $\int \frac{1}{1+e^x}\,dx$.

c Use integration by parts to find $\int x^2 \sin x\,dx$. **E**

8 **a** Find $\int x \sin 2x\,dx$.

b Given that $y = 0$ at $x = \frac{\pi}{4}$, solve the differential equation $\frac{dy}{dx} = x \sin 2x \cos^2 y$. **E**

9 **a** Find $\int x \cos 2x\,dx$.

b This diagram shows part of the curve with equation $y = 2x^{\frac{1}{2}} \sin x$. The shaded region in the diagram is bounded by the curve, the x-axis and the line with equation $x = \frac{\pi}{2}$. This shaded region is rotated through 2π radians about the x-axis to form a solid of revolution. Using calculus, calculate the volume of the solid of revolution formed, giving your answer in terms of π.

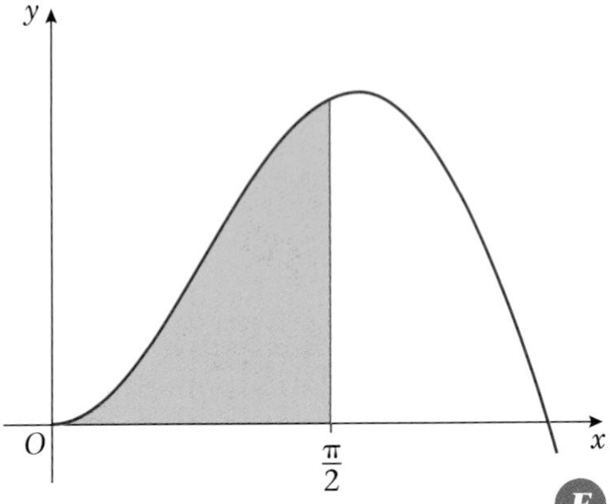

E

10 A curve has equation $y = f(x)$ and passes through the point with coordinates $(0, -1)$. Given that $f'(x) = \frac{1}{2}e^{2x} - 6x$,

a use integration to obtain an expression for $f(x)$,

b show that there is a root α of the equation $f'(x) = 0$, such that $1.41 < \alpha < 1.43$. **E**

11 $f(x) = 16x^{\frac{1}{2}} - \frac{2}{x}$, $x > 0$.

a Solve the equation $f(x) = 0$.

b Find $\int f(x)\,dx$.

c Evaluate $\int_1^4 f(x)\,dx$, giving your answer in the form $p + q \ln r$, where p, q and r are rational numbers. **E**

12

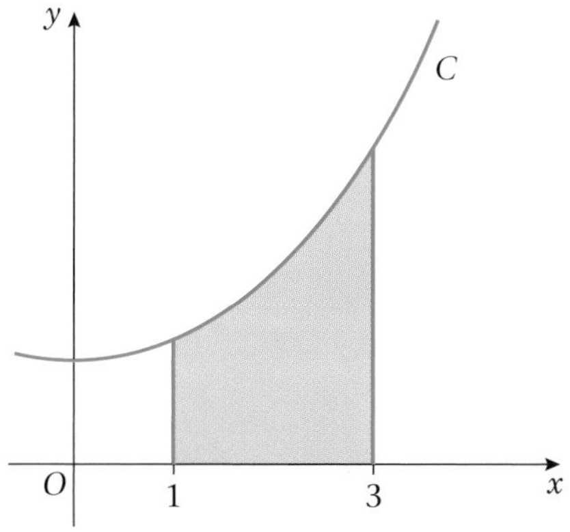

Shown is part of a curve C with equation $y = x^2 + 3$. The shaded region is bounded by C, the x-axis and the lines with equations $x = 1$ and $x = 3$. The shaded region is rotated through 360° about the x-axis.

Using calculus, calculate the volume of the solid generated. Give your answer as an exact multiple of π. **E**

13 **a** Find $\int x(x^2 + 3)^5 \, dx$

b Show that $\int_1^e \frac{1}{x^2} \ln x \, dx = 1 - \frac{2}{e}$

c Given that $p > 1$, show that $\int_1^p \frac{1}{(x+1)(2x-1)} \, dx = \frac{1}{3} \ln \frac{4p-2}{p+1}$ **E**

14 $f(x) \equiv \frac{5x^2 - 8x + 1}{2x(x-1)^2} \equiv \frac{A}{x} + \frac{B}{x-1} + \frac{C}{(x-1)^2}$

a Find the values of the constants A, B and C.

b Hence find $\int f(x) \, dx$.

c Hence show that $\int_4^9 f(x) \, dx = \ln(\frac{32}{3}) - \frac{5}{24}$ **E**

15 The curve shown has parametric equations

$$x = 5 \cos \theta, \; y = 4 \sin \theta, \; 0 \leqslant \theta < 2\pi.$$

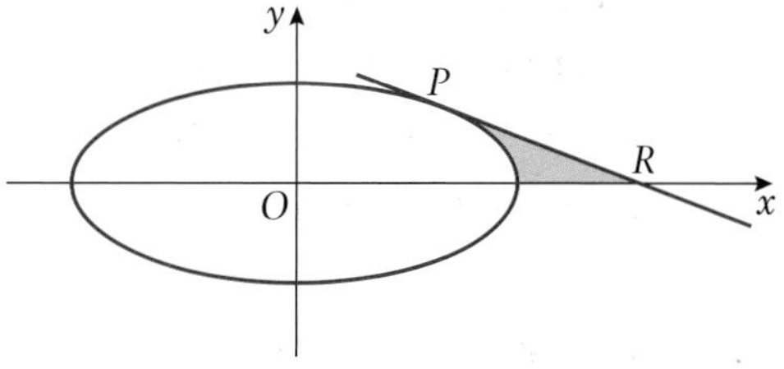

a Find the gradient of the curve at the point P at which $\theta = \frac{\pi}{4}$.

b Find an equation of the tangent to the curve at the point P.

c Find the coordinates of the point R where this tangent meets the x-axis.

The shaded region is bounded by the tangent PR, the curve and the x-axis.

d Find the area of the shaded region, leaving your answer in terms of π. **E**

16 **a** Obtain the general solution of the differential equation

$$\frac{dy}{dx} = xy^2, \; y > 0.$$

b Given also that $y = 1$ at $x = 1$, show that

$$y = \frac{2}{3 - x^2}, \; -\sqrt{3} < x < \sqrt{3}$$

is a particular solution of the differential equation.

The curve C has equation $y = \frac{2}{3 - x^2}$, $x \neq -\sqrt{3}$, $x \neq \sqrt{3}$

c Write down the gradient of C at the point $(1, 1)$.

d Deduce that the line which is a tangent to C at the point $(1, 1)$ has equation $y = x$.

e Find the coordinates of the point where the line $y = x$ again meets the curve C. **E**

17

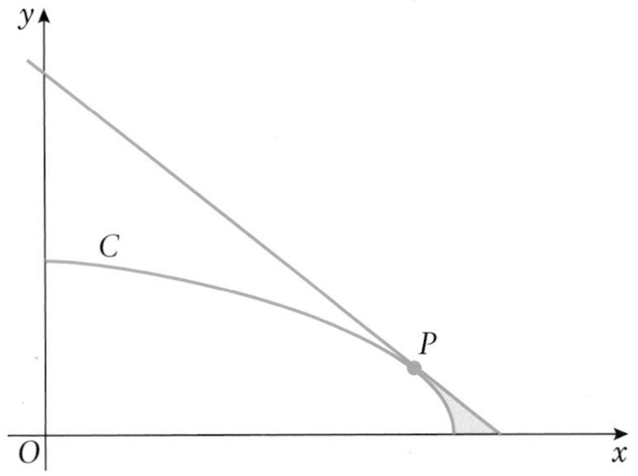

The diagram shows the curve C with parametric equations

$$x = a\sin^2 t, \; y = a\cos t, \; 0 \leqslant t \leqslant \tfrac{1}{2}\pi,$$

where a is a positive constant. The point P lies on C and has coordinates $(\tfrac{3}{4}a, \tfrac{1}{2}a)$.

a Find $\frac{dy}{dx}$, giving your answer in terms of t.

b Find an equation of the tangent to C at P.

c Show that a cartesian equation of C is $y^2 = a^2 - ax$.

The shaded region is bounded by C, the tangent at P and the x-axis. This shaded region is rotated through 2π radians about the x-axis to form a solid of revolution.

d Use calculus to calculate the volume of the solid revolution formed, giving your answer in the form $k\pi a^3$, where k is an exact fraction. **E**

18 **a** Using the substitution $u = 1 + 2x$, or otherwise, find

$$\int \frac{4x}{(1 + 2x)^2}\, dx, \; x > -\tfrac{1}{2},$$

b Given that $y = \frac{\pi}{4}$ when $x = 0$, solve the differential equation

$$(1 + 2x)^2 \frac{dy}{dx} = \frac{x}{\sin^2 y}$$ **E**

19 The diagram shows the curve with equation $y = xe^{2x}$, $-\frac{1}{2} \leqslant x \leqslant \frac{1}{2}$.

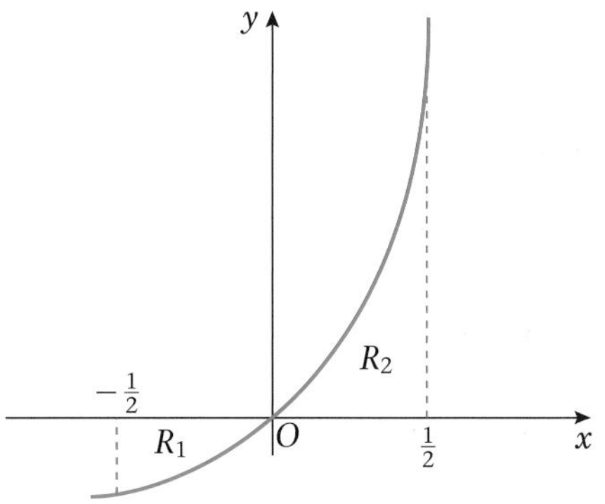

The finite region R_1 bounded by the curve, the x-axis and the line $x = -\frac{1}{2}$ has area A_1.
The finite region R_2 bounded by the curve, the x-axis and the line $x = \frac{1}{2}$ has area A_2.

a Find the exact values of A_1 and A_2 by integration.

b Show that $A_1 : A_2 = (\mathrm{e} - 2) : \mathrm{e}$. **E**

20 Find $\int x^2 \mathrm{e}^{-x}\,\mathrm{d}x$.

Given that $y = 0$ at $x = 0$, solve the differential equation $\frac{\mathrm{d}y}{\mathrm{d}x} = x^2 \mathrm{e}^{3y - x}$. **E**

21 The curve with equation $y = \mathrm{e}^{3x} + 1$ meets the line $y = 8$ at the point $(h, 8)$.

a Find h, giving your answer in terms of natural logarithms.

b Show that the area of the finite region enclosed by the curve with equation $y = \mathrm{e}^{3x} + 1$, the x-axis, the y-axis and the line $x = h$ is $2 + \frac{1}{3}\ln 7$. **E**

22 **a** Given that

$$\frac{x^2}{x^2 - 1} \equiv A + \frac{B}{x - 1} + \frac{C}{x + 1},$$

find the values of the constants A, B and C.

b Given that $x = 2$ at $t = 1$, solve the differential equation

$$\frac{\mathrm{d}x}{\mathrm{d}t} = 2 - \frac{2}{x^2},\ x > 1.$$

You need not simplify your final answer. **E**

23 The curve C is given by the equations

$$x = 2t,\ y = t^2,$$

where t is a parameter.

a Find an equation of the normal to C at the point P on C where $t = 3$.

The normal meets the y-axis at the point B. The finite region R is bounded by the part of the curve C between the origin O and P, and the lines OB and OP.

b Show the region R, together with its boundaries, in a sketch.

The region R is rotated through 2π about the y-axis to form a solid S.

c Using integration, and explaining each step in your method, find the volume of S, giving your answer in terms of π. **E**

24

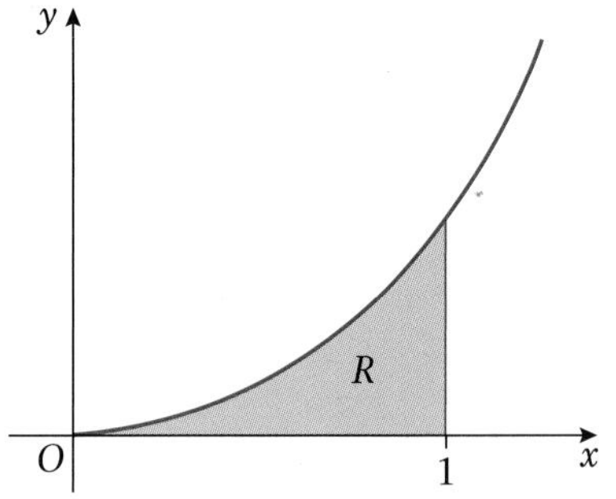

Shown is part of the curve with equation $y = e^{2x} - e^{-x}$. The shaded region R is bounded by the curve, the x-axis and the line with equation $x = 1$.

Use calculus to find the area of R, giving your answer in terms of e. **E**

25 **a** Given that $2y = x - \sin x \cos x$, show that $\frac{dy}{dx} = \sin^2 x$.

b Hence find $\int \sin^2 x \, dx$.

c Hence, using integration by parts, find $\int x \sin^2 x \, dx$. **E**

26 The rate, in $cm^3 \, s^{-1}$, at which oil is leaking from an engine sump at any time t seconds is proportional to the volume of oil, $V \, cm^3$, in the sump at that instant. At time $t = 0$, $V = A$.

a By forming and integrating a differential equation, show that

$$V = Ae^{-kt}$$

where k is a positive constant.

b Sketch a graph to show the relation between V and t.

Given further that $V = \frac{1}{2}A$ at $t = T$,

c show that $kT = \ln 2$. **E**

27

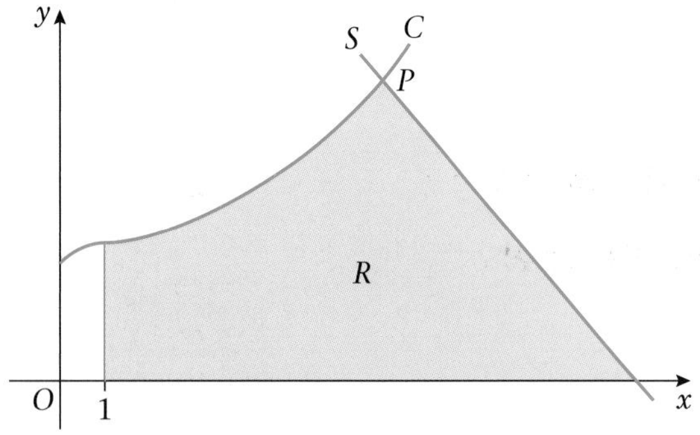

This graph shows part of the curve C with parametric equations

$$x = (t + 1)^2, \; y = \tfrac{1}{2}t^3 + 3, \; t \geqslant -1.$$

P is the point on the curve where $t = 2$. The line S is the normal to C at P.

a Find an equation of S.

The shaded region R is bounded by C, S, the x-axis and the line with equation $x = 1$.

b Using integration and showing all your working, find the area of R. **E**

28

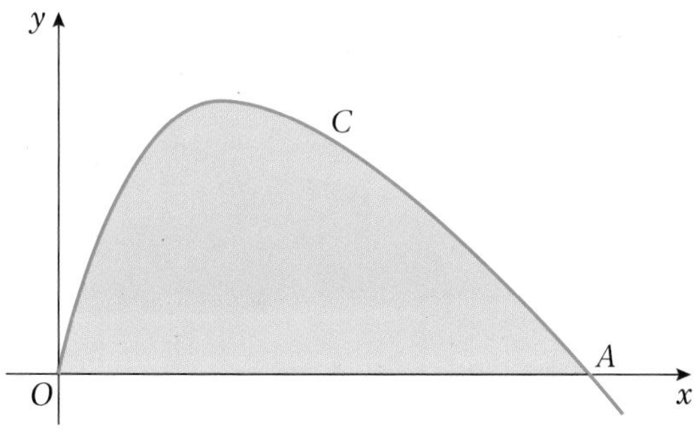

Shown is part of the curve C with parametric equations

$$x = t^2,\ y = \sin 2t,\ t \geqslant 0.$$

The point A is an intersection of C with the x-axis.

a Find, in terms of π, the x-coordinate of A.

b Find $\frac{dy}{dx}$ in terms of t, $t > 0$.

c Show that an equation of the tangent to C at A is $4x + 2\pi y = \pi^2$.

The shaded region is bounded by C and the x-axis.

d Use calculus to find, in terms of π, the area of the shaded region. **E**

29 Showing your method clearly in each case, find

a $\int \sin^2 x \cos x \, dx$,

b $\int x \ln x \, dx$.

Using the substitution $t^2 = x + 1$, where $x > -1$, $t > 0$,

c Find $\int \frac{x}{\sqrt{x+1}} \, dx$.

d Hence evaluate $\int_0^3 \frac{x}{\sqrt{x+1}} \, dx$. **E**

30 **a** Using the substitution $u = 1 + 2x^2$, find $\int x(1 + 2x^2)^5 \, dx$.

b Given that $y = \frac{\pi}{8}$ at $x = 0$, solve the differential equation

$$\frac{dy}{dx} = x(1 + 2x^2)^5 \cos^2 2y.$$ **E**

31 Find $\int x^2 \ln 2x \, dx$. **E**

32 Obtain the solution of

$$x(x + 2)\frac{dy}{dx} = y,\ y > 0,\ x > 0,$$

for which $y = 2$ at $x = 2$, giving your answer in the form $y^2 = f(x)$. **E**

33 **a** Use integration by parts to show that

$$\int_0^{\frac{\pi}{4}} x \sec^2 x \, dx = \tfrac{1}{4}\pi - \tfrac{1}{2}\ln 2.$$

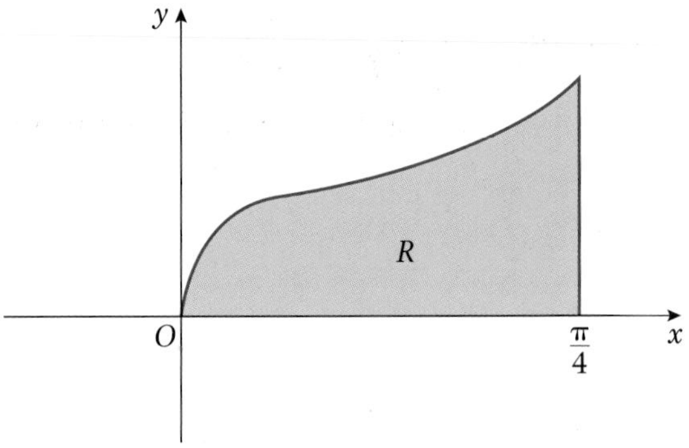

The finite region R, bounded by the curve with equation $y = x^{\frac{1}{2}} \sec x$, the line $x = \frac{\pi}{4}$ and the x-axis is shown. The region R is rotated through 2π radians about the x-axis.

b Find the volume of the solid of revolution generated.

c Find the gradient of the curve with equation $y = x^{\frac{1}{2}} \sec x$ at the point where $x = \frac{\pi}{4}$. **E**

34 Part of the design of a stained glass window is shown. The two loops enclose an area of blue glass. The remaining area within the rectangle $ABCD$ is red glass.

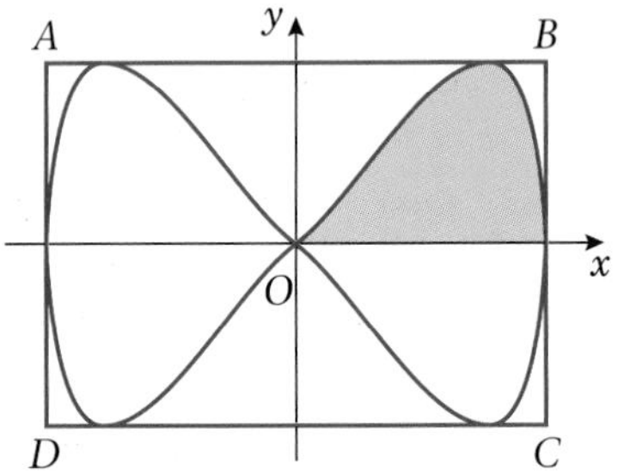

The loops are described by the curve with parametric equations

$$x = 3\cos t,\ y = 9\sin 2t,\ 0 \leqslant t < 2\pi.$$

a Find the cartesian equation of the curve in the form $y^2 = f(x)$.

b Show that the shaded area enclosed by the curve and the x-axis, is given by

$$\int_0^{\frac{\pi}{2}} A \sin 2t \sin t \, dt,$$ stating the value of the constant A.

c Find the value of this integral.

The sides of the rectangle $ABCD$ are the tangents to the curve that are parallel to the coordinate axes. Given that 1 unit on each axis represents 1 cm,

d find the total area of the red glass. **E**

Summary of key points

1 You should be familiar with the following integrals.

$$\int x^n = \frac{x^{n+1}}{n+1} + C$$

$$\int e^x = e^x + C$$

$$\int \frac{1}{x} = \ln|x| + C$$

$$\int \cos x = \sin x + C$$

$$\int \sin x = -\cos x + C$$

$$\int \sec^2 x = \tan x + C$$

$$\int \operatorname{cosec} x \cot x = -\operatorname{cosec} x + C$$

$$\int \operatorname{cosec}^2 x = -\cot x + C$$

$$\int \sec x \tan x = \sec x + C$$

2 Using the chain rule in reverse you can obtain generalisations of the above formulae.

$$\int f'(ax+b)\,dx = \frac{1}{a}f(ax+b) + C$$

$$\int (ax+b)^n\,dx = \frac{1}{a}\frac{(ax+b)^{n+1}}{n+1} + C$$

$$\int e^{ax+b}\,dx = \frac{1}{a}e^{ax+b} + C$$

$$\int \frac{1}{ax+b}\,dx = \frac{1}{a}\ln|ax+b| + C$$

$$\int \cos(ax+b)\,dx = \frac{1}{a}\sin(ax+b) + C$$

$$\int \sin(ax+b)\,dx = -\frac{1}{a}\cos(ax+b) + C$$

$$\int \sec^2(ax+b)\,dx = \frac{1}{a}\tan(ax+b) + C$$

$$\int \operatorname{cosec}(ax+b)\cot(ax+b)\,dx = -\frac{1}{a}\operatorname{cosec}(ax+b) + C$$

$$\int \operatorname{cosec}^2(ax+b)\,dx = -\frac{1}{a}\cot(ax+b) + C$$

$$\int \sec(ax+b)\tan(ax+b)\,dx = \frac{1}{a}\sec(ax+b) + C$$

3 Sometimes trigonometric identities can be useful to help change the expression into one you know how to integrate.
e.g. To integrate $\sin^2 x$ or $\cos^2 x$ use formulae for $\cos 2x$, so

$$\int \sin^2 x \, dx = \int (\tfrac{1}{2} - \tfrac{1}{2}\cos 2x) \, dx$$

4 You can use partial fractions to integrate expressions of the type $\dfrac{x-5}{(x+1)(x-2)}$.

5 You should remember the following general patterns:

$$\int \frac{f'(x)}{f(x)} \, dx = \ln|f(x)| + C$$

$$\int f'(x)[f(x)]^n \, dx = \frac{1}{n+1}[f(x)]^{n+1}; \; n \neq -1$$

6 Sometimes you can simplify an integral by changing the variable. This process is similar to using the chain rule in differentiation and is called **integration by substitution**.

7 **Integration by parts:**

$$\int u \frac{dv}{dx} \, dx = uv - \int v \frac{du}{dx} \, dx$$

8 $\int \tan x \, dx = \ln|\sec x| + C$

$$\int \sec x \, dx = \ln|\sec x + \tan x| + C$$

$$\int \cot x \, dx = \ln|\sin x| + C$$

$$\int \operatorname{cosec} x \, dx = -\ln|\operatorname{cosec} x + \cot x| + C$$

9 Remember: the trapezium rule is

$$\int_a^b y \, dx \approx \tfrac{1}{2}h[y_0 + 2(y_1 + y_2 + \ldots + y_{n-1}) + y_n]$$

where $h = \dfrac{b-a}{n}$ and $y_i = f(a + ih)$

10 Area of region between $y = f(x)$, the x-axis and $x = a$ and $x = b$ is given by:

$$\text{Area} = \int_a^b y \, dx$$

11 Volume of revolution formed by rotating y about the x-axis between $x = a$ and $x = b$ is given by:

$$\text{Volume} = \pi \int_a^b y^2 \, dx$$

12 When $\dfrac{dy}{dx} = f(x)g(y)$ you can write

$$\int \frac{1}{g(y)} \, dy = \int f(x) \, dx$$

This is called separating the variables.

Examination style paper

1 Use the binomial theorem to expand $\frac{1}{(2+x)^2}$, $|x| < 2$, in ascending powers of x, as far as the term in x^3, giving each coefficient as a simplified fraction. (6)

2 The curve C has equation

$$x^2 + 2y^2 - 4x - 6yx + 3 = 0$$

Find the gradient of C at the point (1, 3). (7)

3 Use the substitution $u = 5x + 3$, to find an exact value for

$$\int_0^3 \frac{10x}{(5x+3)^3}\,dx$$ (9)

4 a Find the values of A and B for which

$$\frac{1}{(2x+1)(x-2)} \equiv \frac{A}{2x+1} + \frac{B}{x-2}$$ (3)

b Hence find $\int \frac{1}{(2x+1)(x-2)}\,dx$, giving your answer in the form $\ln f(x)$. (4)

c Hence, or otherwise, obtain the solution of

$$(2x+1)(x-2)\frac{dy}{dx} = 10y,\ y > 0,\ x > 2$$

for which $y = 1$ at $x = 3$, giving your answer in the form $y = f(x)$. (5)

5 A population grows in such a way that the rate of change of the population P at time t in days is proportional to P.

a Write down a differential equation relating P and t. (2)

b Show, by solving this equation or by differentiation, that the general solution of this equation may be written as $P = Ak^t$, where A and k are positive constants. (5)

Initially the population is 8 million and 7 days later it has grown to 8.5 million.

c Find the size of the population after a further 28 days. (5)

6 Referred to an origin O the points A and B have position vectors $\mathbf{i} - 5\mathbf{j} - 7\mathbf{k}$ and $10\mathbf{i} + 10\mathbf{j} + 5\mathbf{k}$ respectively. P is a point on the line AB.

a Find a vector equation for the line passing through A and B. (3)

b Find the position vector of point P such that OP is perpendicular to AB. (5)

c Find the area of triangle OAB. (4)

d Find the ratio in which P divides the line AB. (2)

7

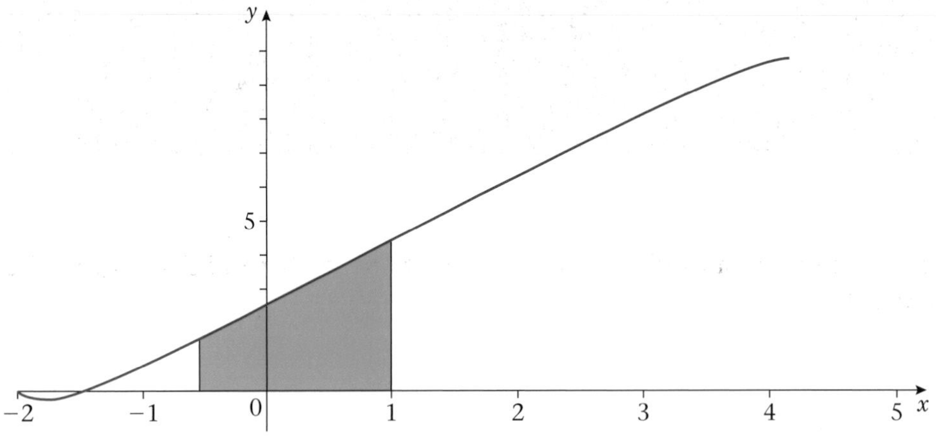

The curve C, shown has parametric equations

$$x = 1 - 3\cos t,\ y = 3t - 2\sin 2t,\ 0 < t < \pi.$$

a Find the gradient of the curve at the point P where $t = \dfrac{\pi}{6}$. (4)

b Show that the area of the finite region beneath the curve, between the lines $x = -\frac{1}{2}$, $x = 1$ and the x-axis, shown shaded in the diagram, is given by the integral

$$\int_{\frac{\pi}{3}}^{\frac{\pi}{2}} 9t \sin t \, dt - \int_{\frac{\pi}{3}}^{\frac{\pi}{2}} 12 \sin^2 t \cos t \, dt.$$ (4)

c Hence, by integration, find an exact value for this area. (7)

Formulae you need to remember

This appendix lists formulae that candidates are expected to remember and that may not be included in formulae booklets.

Integration

function	integral
$\cos kx$	$\frac{1}{k}\sin kx + c$
e^{kx}	$\frac{1}{k}e^{kx} + c$
$\frac{1}{x}$	$\ln \lvert x \rvert + c, x \neq 0$
$f'(x) + g'(x)$	$f'(x) + g'(x) + c$
$f'(g(x))g'(x)$	$f(g(x)) + c$

Vectors

$$\begin{pmatrix} x \\ y \\ z \end{pmatrix} . \begin{pmatrix} a \\ b \\ c \end{pmatrix} = xa + yb + zc$$

List of symbols and notation

The following notation will be used in all Edexcel mathematics examinations:

$\in$	is an element of
$\notin$	is not an element of
$\{x_1, x_2, \ldots\}$	the set with elements $x_1, x_2, \ldots$
$\{x: \ldots\}$	the set of all x such that …
$\text{n}(A)$	the number of elements in set A
$\varnothing$	the empty set
ξ	the universal set
A'	the complement of the set A
$\mathbb{N}$	the set of natural numbers, $\{1, 2, 3, \ldots\}$
$\mathbb{Z}$	the set of integers, $\{0, \pm1, \pm2, \pm3, \ldots\}$
$\mathbb{Z}^+$	the set of positive integers, $\{1, 2, 3, \ldots\}$
$\mathbb{Z}_n$	the set of integers modulo n, $\{1, 2, 3, \ldots, n-1\}$
$\mathbb{Q}$	the set of rational numbers, $\left\{\frac{p}{q}: p \in \mathbb{Z}_w, q \in \mathbb{Z}^+\right\}$
$\mathbb{Q}^+$	the set of positive rational numbers, $\{x \in \mathbb{Q}: x > 0\}$
$\mathbb{Q}_0^+$	the set of positive rational numbers and zero, $\{x \in \mathbb{Q}: x \geqslant 0\}$
$\mathbb{R}$	the set of real numbers
$\mathbb{R}^+$	the set of positive real numbers, $\{x \in \mathbb{R}: x > 0\}$
$\mathbb{R}_0^+$	the set of positive real numbers and zero, $\{x \in \mathbb{R}: x \geqslant 0\}$
$\mathbb{C}$	the set complex numbers
(x, y)	the ordered pair x, y
$A \times B$	the cartesian products of sets A and B, ie $A \times B = \{(a, b): a \in A, b \in B\}$
$\subseteq$	is a subset of
$\subset$	is a proper subset of
$\cup$	union
$\cap$	intersection
$[a, b]$	the closed interval, $\{x \in \mathbb{R}: a \leqslant x \leqslant b\}$
$[a, b), [a, b[$	the interval, $\{x \in \mathbb{R}: a \leqslant x < b\}$
$(a, b],]a, b]$	the interval, $\{x \in \mathbb{R}: a < x \leqslant b\}$
$(a, b),]a, b[$	the open interval, $\{x \in \mathbb{R}: a < x < b\}$
$y R x$	y is related to x by the relation R
$y \sim x$	y is equivalent to x, in the context of some equivalence relation
$=$	is equal to
$\neq$	is not equal to
$\equiv$	is identical to or is congruent to
$\approx$	is approximately equal to
$\cong$	is isomorphic to
$\propto$	is proportional to
$<$	is less than
$\leqslant, \not>$	is less than or equal to, is not greater than

$>$	is greater than
$\geqslant$, $\nless$	is greater than or equal to, is not less than
∞	infinity
$p \wedge q$	p and q
$p \vee q$	p or q (or both)
$\sim p$	not p
$p \Rightarrow q$	p implies q (if p then q)
$p \Leftarrow q$	p is implied by q (if q then p)
$p \Leftrightarrow q$	p implies and is implied by q (p is equivalent to q)
$\exists$	there exists
$\forall$	for all
$a + b$	a plus b
$a - b$	a minus b
$a \times b$, ab, $a.b$	a multiplied by b
$a \div b$, $\frac{a}{b}$, a/b	a divided by b
$\sum_{i=1}^{n}$	$a_1 + a_2 + \ldots + a_n$
$\prod_{i=1}^{n}$	$a_1 \times a_2 \times \ldots \times a_n$
$\sqrt{a}$	the positive square root of a
$\lvert a \rvert$	the modulus of a
$n!$	n factorial
$\binom{n}{r}$	the binomial coefficient $\frac{n!}{r!(n-r)!}$ for $n \in \mathbb{Z}^+$
	$\frac{n(n-1)\ldots(n-r+1)}{r!}$ for $n \in \mathbb{Q}$
$\mathrm{f}(x)$	the value of the function f at x
$\mathrm{f}: A \to B$	f is a function under which each element of set A has an image in set B
$\mathrm{f}: x \to y$	the function f maps the element x to the element y
f^{-1}	the inverse function of the function f
$\mathrm{g} \circ \mathrm{f}$, gf	the composite function of f and g which is defined by $(\mathrm{g} \circ \mathrm{f})(x)$ or $\mathrm{gf}(x) = \mathrm{g}(\mathrm{f}(x))$
$\lim_{x \to a} \mathrm{f}(x)$	the limit of(x) of as x tends to a
Δx, δx	an increment of x
$\frac{\mathrm{d}y}{\mathrm{d}x}$	the derivative of y with respect to x
$\frac{\mathrm{d}^n y}{\mathrm{d}x^n}$	the nth derivative of y with respect to x
$\mathrm{f}'(x), \mathrm{f}''(x), \ldots, \mathrm{f}^{(n)}(x)$	the first, second, ..., nth derivatives of $\mathrm{f}(x)$ with respect to x
$\int y \, \mathrm{d}x$	the indefinite integral of y with respect to x
$\int_b^a y \, \mathrm{d}x$	the definite integral of y with respect to x between the limits
$\frac{\partial V}{\partial x}$	the partial derivative of V with respect to x
$\dot{x}, \ddot{x}, \ldots$	the first, second, ... derivatives of x with respect to t

e	base of natural logarithms
e^x, $\exp x$	exponential function of x
$\log_a x$	logarithm to the base a of x
$\ln x$, $\log_e x$	natural logarithm of x
$\lg x$, $\log_{10} x$	logarithm of x to base 10
sin, cos, tan, cosec, sec, cot	the circular functions
arcsin, arccos, arctan, arccosec, arcsec, arccot	the inverse circular functions
sinh, cosh, tanh, cosech, sech, coth	the hyperbolic functions
arsinh, arcosh, artanh, arcosech, arsech, arcoth	the inverse hyperbolic functions
i, j	square root of -1
z	a complex number, $z = x + \mathrm{i}y$
$\mathrm{Re}\, z$	the real part of z, $\mathrm{Re}\, z = x$
$\mathrm{Im}\, z$	the imaginary part of z, $\mathrm{Im}\, z = y$
$\lvert z \rvert$	the modulus of z, $\lvert z \rvert = \sqrt{(x^2 + y^2)}$
$\arg z$	the argument of z, $\arg z = \theta$, $-\pi < \theta \leqslant \pi$
z^*	the complex conjugate of z, $x - \mathrm{i}y$
$\mathbf{M}$	a matrix $\mathbf{M}$
$\mathbf{M}^{-1}$	the inverse of the matrix $\mathbf{M}$
$\mathbf{M}^{\mathrm{T}}$	the transpose of the matrix $\mathbf{M}$
$\det \mathbf{M}$ or $\lvert \mathbf{M} \rvert$	the determinant of the square matrix $\mathbf{M}$
$\mathbf{a}$	the vector $\mathbf{a}$
$\overrightarrow{AB}$	the vector represented in magnitude and direction by the directed line segment AB
$\hat{\mathbf{a}}$	a unit vector in the direction of $\mathbf{a}$
$\mathbf{i}, \mathbf{j}, \mathbf{k}$	unit vectors in the direction of the cartesian coordinate axes
$\lvert \mathbf{a} \rvert$, a	the magnitude of $\mathbf{a}$
$\lvert \overrightarrow{AB} \rvert$	the magnitude of $\overrightarrow{AB}$
$\mathbf{a} . \mathbf{b}$	the scalar product of $\mathbf{a}$ and $\mathbf{b}$
$\mathbf{a} \times \mathbf{b}$	the vector product of $\mathbf{a}$ and $\mathbf{b}$

Answers

Exercise 1A

1 $\frac{7}{12}$

2 $\frac{7}{20}$

3 $\dfrac{x+3}{x(x+1)}$

4 $\dfrac{5x+1}{(x-1)(x+2)}$

5 $\dfrac{8x-2}{(2x+1)(x-1)}$

6 $\dfrac{5x+34}{(x-3)(x+4)}$

7 $\dfrac{-9x-3}{2x(x-1)}$

8 $\dfrac{6x^2+14x+6}{x(x+1)(x+2)}$

9 $\dfrac{-x^2-24x-8}{3x(x-2)(2x+1)}$

10 $\dfrac{9x^2-14x-7}{(x-1)(x+1)(x-3)}$

Exercise 1B

1 a $\dfrac{2}{(x-2)}+\dfrac{4}{(x+3)}$

b $\dfrac{3}{(x+1)}-\dfrac{1}{(x+4)}$

c $\dfrac{3}{2x}-\dfrac{5}{(x-4)}$

d $\dfrac{4}{(2x+1)}-\dfrac{1}{(x-3)}$

e $\dfrac{2}{(x+3)}+\dfrac{4}{(x-3)}$

f $-\dfrac{1}{(x-4)}-\dfrac{2}{(x+1)}$

g $\dfrac{2}{x}-\dfrac{3}{(x+4)}$

h $-\dfrac{1}{(x-3)}+\dfrac{3}{(x+5)}$

2 $A=\frac{1}{2}$, $B=-\frac{3}{2}$

Exercise 1C

1 a $\dfrac{1}{(x+1)}-\dfrac{2}{(x-2)}+\dfrac{3}{(x+5)}$

b $-\dfrac{1}{x}+\dfrac{2}{(2x+1)}-\dfrac{5}{(3x-2)}$

c $\dfrac{3}{(x+1)}-\dfrac{2}{(x+2)}-\dfrac{6}{(x-5)}$

2 a $\dfrac{3}{x}-\dfrac{2}{(x+1)}+\dfrac{5}{(x-1)}$

b $\dfrac{4}{x}+\dfrac{2}{(x+1)}-\dfrac{1}{(x+2)}$

c $\dfrac{6}{(x-2)}-\dfrac{2}{(x-3)}+\dfrac{1}{(x+1)}$

Exercise 1D

1 $\dfrac{2}{x^2}-\dfrac{1}{x}+\dfrac{4}{(x+1)}$

2 $\dfrac{3}{(x+1)}-\dfrac{2}{(x+1)^2}-\dfrac{4}{(x-1)}$

3 $-\dfrac{2}{x}+\dfrac{4}{(x-3)}+\dfrac{2}{(x-3)^2}$

4 $\dfrac{4}{x}+\dfrac{3}{(x-4)}+\dfrac{2}{(x-4)^2}$

5 $\dfrac{3}{x}+\dfrac{1}{x^2}+\dfrac{2}{(x-1)}$

6 $-\dfrac{2}{x}+\dfrac{4}{(x-3)}+\dfrac{2}{(x-3)^2}$

7 $\dfrac{2}{(x+2)}-\dfrac{4}{(x+2)^2}$

8 $\dfrac{1}{(x+2)}+\dfrac{1}{(x+2)^2}+\dfrac{1}{(x+2)^3}$

Exercise 1E

1 a $1+\dfrac{1}{(x+1)}+\dfrac{4}{(x-3)}$

b $1-\dfrac{2}{(x-2)}+\dfrac{3}{(x+1)}$

c $x+\dfrac{3}{x}-\dfrac{4}{(x-1)}$

d $2-\dfrac{4}{(x+1)}+\dfrac{1}{(x+1)^2}$

2 a $4+\dfrac{2}{(x-1)}+\dfrac{3}{(x+4)}$

b $x+\dfrac{3}{x}+\dfrac{2}{(x-2)}-\dfrac{1}{(x-2)^2}$

3 $A=2$, $B=-3$, $C=5$, $D=1$

Mixed exercise 1F

1 a $\dfrac{3}{x}-\dfrac{2}{(x-1)}$

b $\dfrac{4}{x}-\dfrac{2}{x^2}+\dfrac{3}{(x+1)}$

c $\dfrac{1}{(x-2)}+\dfrac{2}{(x+1)}-\dfrac{3}{(x-5)}$

d $1-\dfrac{1}{2x}+\dfrac{5}{2(x-2)}$

2 a $\dfrac{3}{(x+1)}-\dfrac{2}{(x+1)^2}$

b $2-\dfrac{1}{(x+3)}-\dfrac{1}{(x-1)}$

c $\dfrac{3}{(x+2)}-\dfrac{4}{(x+2)^3}$

d $x^2+2x+3+\dfrac{1}{(x-1)^2}+\dfrac{4}{(x-1)}$

3 a $f(-3)=0$ or $f(x)=(x+3)(2x^2+3x+1)$

b $\dfrac{1}{(x+3)}+\dfrac{8}{(2x+1)}-\dfrac{5}{(x+1)}$

Exercise 2A

1

t	−5	−4	−3	−2	−1	−0.5	0.5	1	2	3	4	5
$x=2t$	−10	−8	−6	−4	−2	−1	1	2	4	6	8	10
$y=\frac{5}{t}$	−1	−1.25	−1.67	−2.5	−5	−10	10	5	2.5	1.67	1.25	1

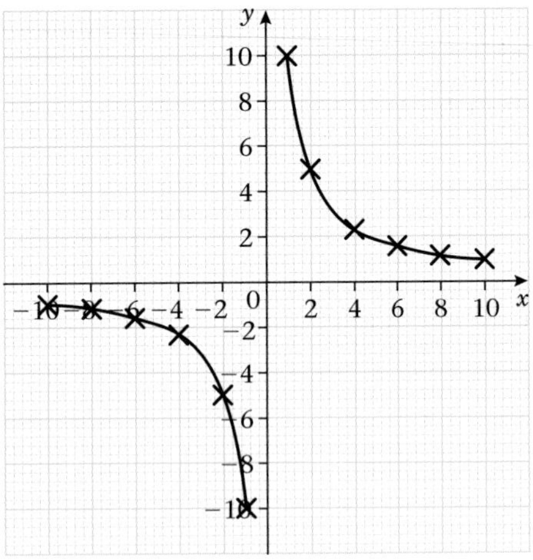

2

t	−4	−3	−2	−1	0	1	2	3	4
$x = t^2$	16	9	4	1	0	1	4	9	16
$y = \frac{t^3}{5}$	−12.8	−5.4	−1.6	−0.2	0	0.2	1.6	5.4	12.8

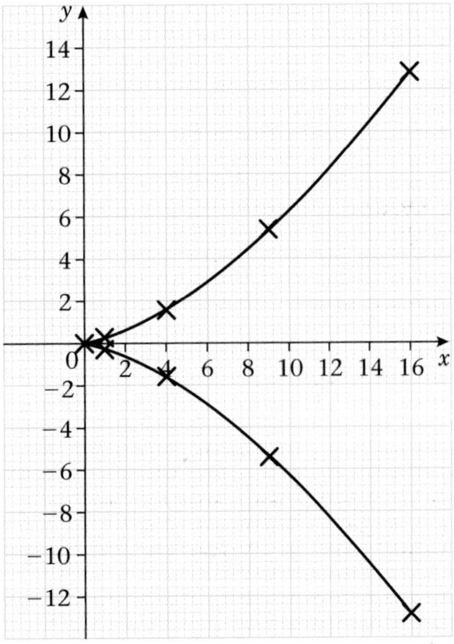

3 a

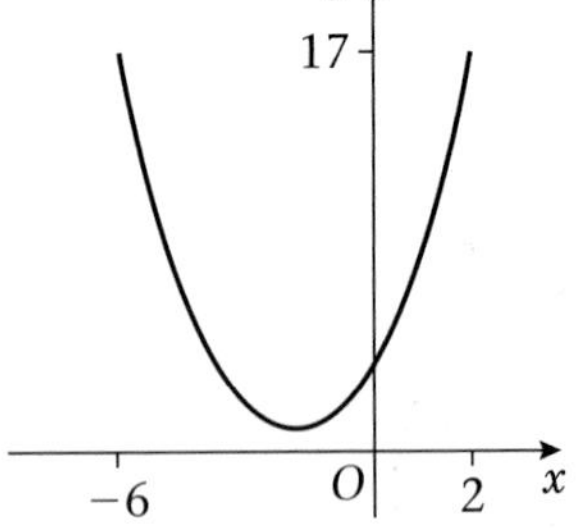

b

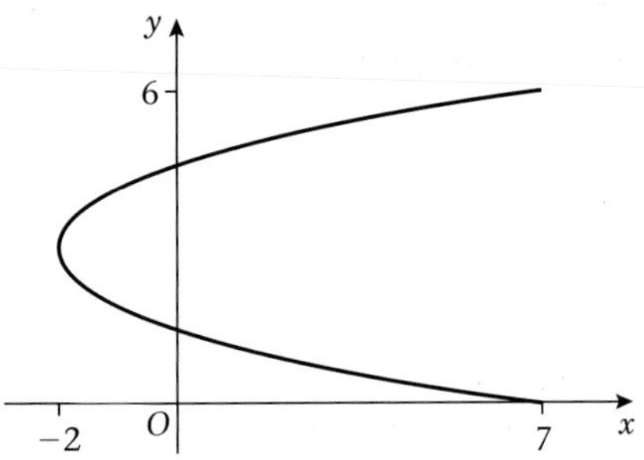

c

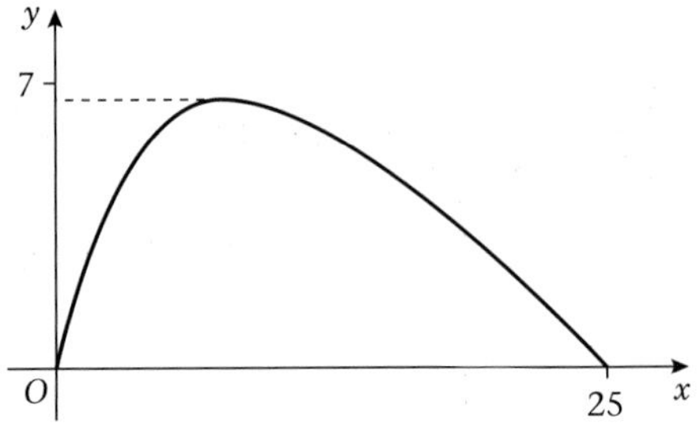

d

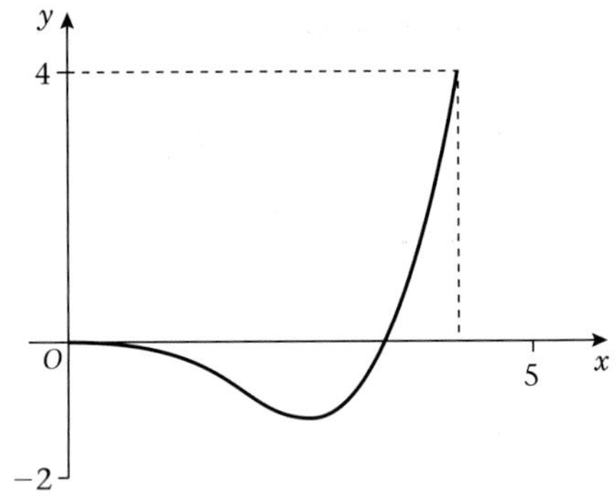

e

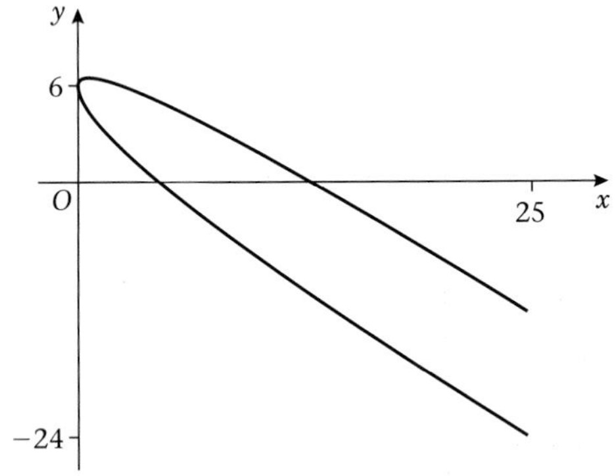

4 **a** $y = (x + 2)^2$ **b** $y = x^2 - 10x + 24$

c $y = 3 - \dfrac{1}{x}$ **d** $y = \dfrac{2}{x - 1}$

e $y = \dfrac{15 - x}{2}$ **f** $y = x^2(9 - x^2)$

g $\frac{1}{9}(x - 2)(x + 7)$ **h** $y = \left(\dfrac{1 + 2x}{x}\right)^2$

i $y = \dfrac{x}{1 - 3x}$ **j** $y = \dfrac{x}{3x - 1}$

5 $y = \frac{3}{2}x + \frac{1}{2}$

Exercise 2B

1 **a** (11, 0) **b** (7, 0) **c** (1, 0), (9, 0)

d (1, 0), (2, 0) **e** $(\frac{9}{5}, 0)$

2 **a** (0, −5) **b** $(0, \frac{9}{16})$ **c** (0, 0), (0, 12)
d $(0, \frac{1}{2})$ **e** (0, 1)

3 4

4 4

5 **a** $p = -\frac{2}{5}$ **b** $(0, -\frac{13}{4})$

6 $\frac{8}{3}$

7 $(\frac{1}{2}, \frac{3}{2})$

8 (1, 0), (17, 12)

9 $t = \frac{5}{2}$, $t = -\frac{3}{2}$; $(\frac{25}{4}, 5)$, $(\frac{9}{4}, -3)$

10 (1, 2), (1, −2), (4, 4), (4, −4)

Exercise 2C

1

t	0	$\frac{\pi}{6}$	$\frac{\pi}{3}$	$\frac{\pi}{2}$	$\frac{2\pi}{3}$	$\frac{5\pi}{6}$	π	$\frac{7\pi}{6}$	$\frac{4\pi}{3}$	$\frac{3\pi}{2}$	$\frac{5\pi}{3}$	$\frac{11\pi}{6}$	2π
$x = 2\sin t$	0	1	1.73	2	1.73	1	0	−1	−1.73	−2	−1.73	−1	0
$y = \cos t$	1	0.87	0.5	0	−0.5	−0.87	−1	−0.87	−0.5	0	0.5	0.87	1

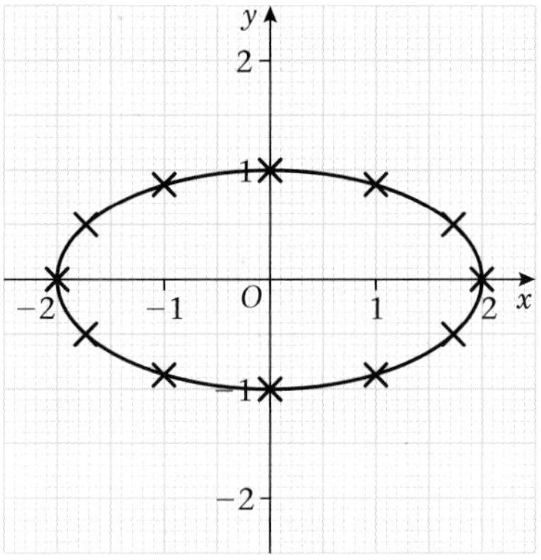

2

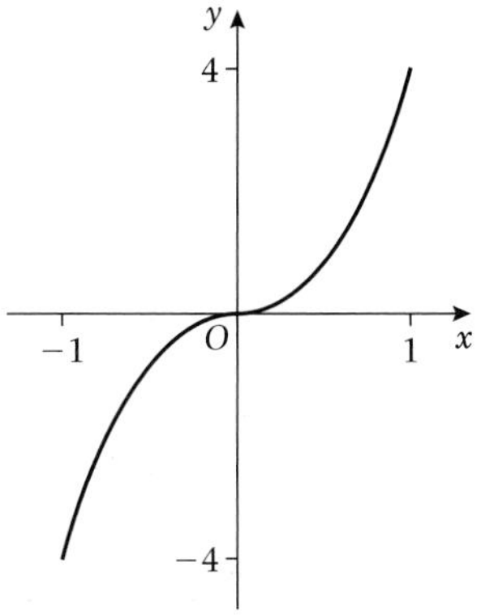

3 **a** $x^2 + y^2 = 1$
b $(x + 3)^2 + y^2 = 1$
c $(x + 2)^2 + (y - 3)^2 = 1$
d $\left(\frac{x}{2}\right)^2 + \left(\frac{y}{3}\right)^2 = 1$
e $\left(\frac{x + 1}{2}\right)^2 + \left(\frac{y - 4}{5}\right)^2 = 1$
f $y = 2x\sqrt{1 - x^2}$
g $y = 4x^2 - 2$
h $y = \frac{x}{\sqrt{1 - x^2}}$
i $y = \frac{4}{x - 2}$
j $y^2 = 1 + \left(\frac{x}{3}\right)^2$

4 **a** $(x + 5)^2 + (y - 2)^2 = 1$ **b** 1, (−5, 2)

5 4, (3, −1)

Exercise 2D

1 **a** $4t - 3$
b $3t^2(t^2 + 1)$
c $4(1 - t^2)(2t - 3)$
d $4t$
e $3t^{\frac{5}{2}}$
f $-\frac{40}{t}$
g $10t^{-2}$
h $\frac{1}{3}t^{-\frac{1}{6}}$
i $8t^2 - 12t^3$
j $4t^{\frac{5}{3}}$

2 112

3 744

4 $306\frac{9}{16}$

5 $17\frac{1}{15}$

6 **a** 12
b 24

7 **a** 2
b 35

8 **a** **i** 2 **ii** 3
b 38

9 $52\frac{1}{12}$

10 **a** **i** −1 **ii** −2 **b** 1

Mixed exercise 2E

1 **a** (4, 0), (0, 3)
b $(2\sqrt{3}, \frac{3}{2})$
c $\left(\frac{x}{4}\right)^2 + \left(\frac{y}{3}\right)^2 = 1$

2

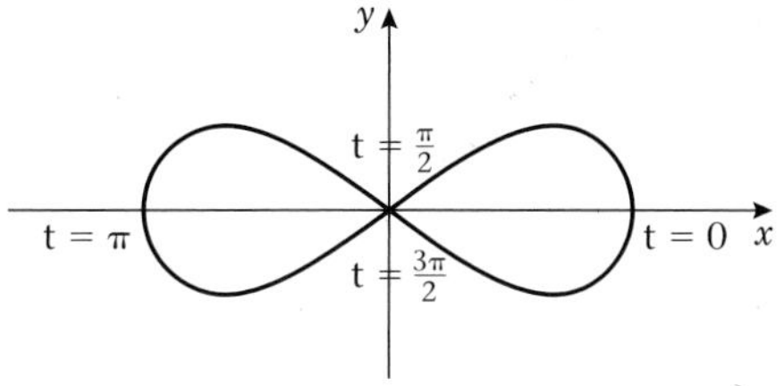

3 **a** $y = 1 - 2x^2$
b $\frac{\sqrt{2}}{2}, -\frac{\sqrt{2}}{2}$

4 $t = \frac{1}{x} - 1$

5 **a** $(x + 3)^2 + (y - 5)^2 = 16$
b

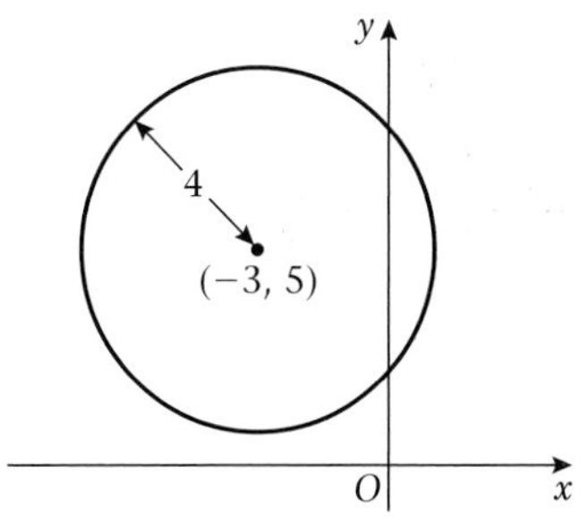

c $(0, 5 + \sqrt{7})$, $(0, 5 - \sqrt{7})$

6 $y = \frac{1}{5}x + \frac{13}{5}$

7 a

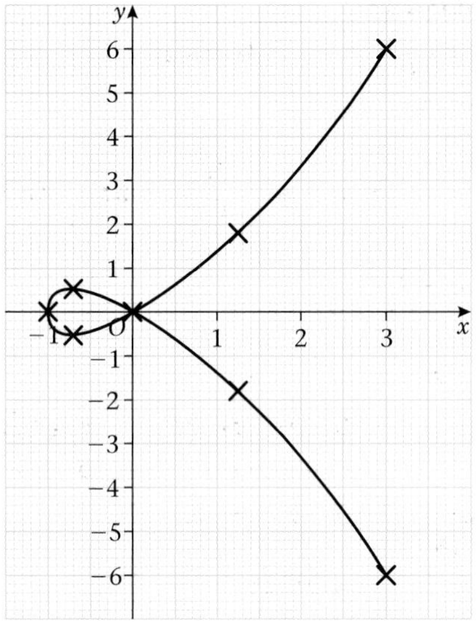

b $\frac{8}{15}$

8 a

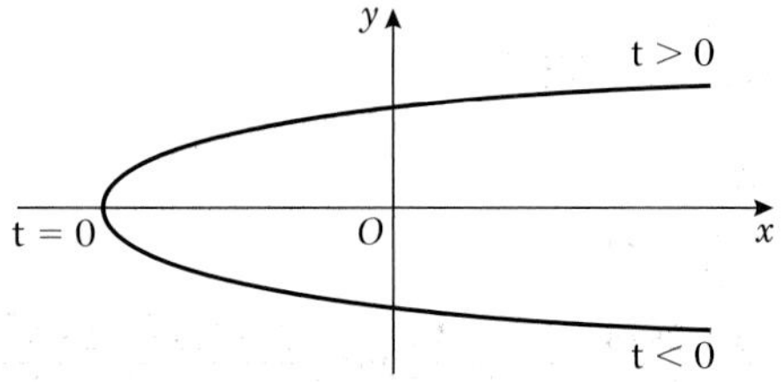

c $\dfrac{16\sqrt{2}}{3}$

9 18

10 b $12\frac{4}{5}$

Exercise 3A

1 a $1 + 6x + 12x^2 + 8x^3$, valid for all x

b $1 + x + x^2 + x^3$, $|x| < 1$

c $1 + \frac{1}{2}x - \frac{1}{8}x^2 + \frac{1}{16}x^3$, $|x| < 1$

d $1 - 6x + 24x^2 - 80x^3$, $|x| < \frac{1}{2}$

e $1 - x - x^2 - \frac{5}{3}x^3$, $|x| < \frac{1}{3}$

f $1 - 15x + \frac{75}{2}x^2 + \frac{125}{2}x^3$, $|x| < \frac{1}{10}$

g $1 - x + \frac{5}{8}x^2 - \frac{5}{16}x^3$, $|x| < 4$

h $1 - 2x^2 + \ldots$, $|x| < \dfrac{\sqrt{2}}{2}$

2 $|x| < \frac{1}{2}$

3 $1 + \dfrac{3x}{2} - \dfrac{9}{8}x^2 + \dfrac{27}{16}x^3$, 10.148 891 88, accurate to 6 d.p.

4 $a = \pm 8$, $\mp 160x^3$

6 $1 - \dfrac{9x}{2} + \dfrac{27x^2}{8} + \dfrac{27x^3}{16}$, $x = 0.01$, 955.339(1875)

Exercise 3B

1 a $2 + \dfrac{x}{2} - \dfrac{x^2}{16} + \dfrac{x^3}{64}$, $|x| < 2$

b $\dfrac{1}{2} - \dfrac{x}{4} + \dfrac{x^2}{8} - \dfrac{x^3}{16}$, $|x| < 2$

c $\dfrac{1}{16} + \dfrac{x}{32} + \dfrac{3x^2}{256} + \dfrac{x^3}{256}$, $|x| < 4$

d $3 + \dfrac{x}{6} - \dfrac{x^2}{216} + \dfrac{x^3}{3888}$, $|x| < 9$

e $\dfrac{\sqrt{2}}{2} - \dfrac{\sqrt{2}}{8}x + \dfrac{3\sqrt{2}}{64}x^2 - \dfrac{5\sqrt{2}}{256}x^3$, $|x| < 2$

f $\dfrac{5}{3} - \dfrac{10}{9}x + \dfrac{20}{27}x^2 - \dfrac{40}{81}x^3$, $|x| < \dfrac{3}{2}$

g $\dfrac{1}{2} + \dfrac{1}{4}x - \dfrac{1}{8}x^2 + \dfrac{1}{16}x^3$, $|x| < 2$

h $\sqrt{2} + \dfrac{3\sqrt{2}}{4}x + \dfrac{15\sqrt{2}}{32}x^2 + \dfrac{51\sqrt{2}}{128}x^3$, $|x| < 1$

2 $|x| < 4$

3 $2 - \dfrac{x}{4} - \dfrac{x^2}{64} - \dfrac{x^3}{512}$, $\dfrac{736\,055}{124\,416}$ accurate to 6 d.p.

4 $a = \pm 2$, $b = \mp 1$, $c = \frac{3}{16}$

Exercise 3C

1 a $\dfrac{4}{(1-x)} - \dfrac{4}{(2+x)}$

b $2 + 5x + \dfrac{7}{2}x^2$

c valid $|x| < 1$

2 a $-\dfrac{2}{(2+x)} + \dfrac{4}{(2+x)^2}$

b $B = \frac{1}{2}$, $C = -\frac{3}{8}$

c $|x| < 2$

3 a $\dfrac{2}{(1+x)} + \dfrac{3}{(1-x)} - \dfrac{4}{(2+x)}$

b $3 + 2x + \dfrac{9}{2}x^2 + \dfrac{5}{4}x^3$

c $|x| < 1$

Mixed exercise 3D

1 a $1 - 12x + 48x^2 - 64x^3$, all x

b $4 + \dfrac{x}{8} - \dfrac{x^2}{512} + \dfrac{x^3}{16\,384}$, $|x| < 16$

c $1 + 2x + 4x^2 + 8x^3$, $|x| < \frac{1}{2}$

d $2 - 3x + \dfrac{9}{2}x^2 - \dfrac{27}{4}x^3$, $|x| < \frac{2}{3}$

e $2 + \dfrac{x}{4} + \dfrac{3}{64}x^2 + \dfrac{5}{512}x^3$, $|x| < 4$

f $1 - 2x + 6x^2 - 18x^3$, $|x| < \frac{1}{3}$

g $1 + 4x + 8x^2 + 12x^3$, $|x| < 1$

h $-3 - 8x - 18x^2 - 38x^3$, $|x| < \frac{1}{2}$

2 $1 - \dfrac{x}{4} - \dfrac{x^2}{32} - \dfrac{x^3}{128}$

4 a $2 - \dfrac{x}{4} - \dfrac{x^2}{64} - \dfrac{x^3}{512}$ **b** $2 + \dfrac{15}{4}x - \dfrac{33}{64}x^2 - \dfrac{17}{512}x^3$

5 a $\dfrac{1}{2} - \dfrac{3}{4}x + \dfrac{9}{8}x^2 - \dfrac{27}{16}x^3$ **b** $\dfrac{1}{2} - \dfrac{x}{4} + \dfrac{3}{8}x^2 - \dfrac{9}{16}x^3$

6 $\dfrac{1}{2} - \dfrac{x}{16} + \dfrac{3}{256}x^2 - \dfrac{5}{2048}x^3$

7 a $1 - 3x + 9x^2 - 27x^3$

b $1 - 2x + 6x^2 - 18x^3$

c $x = 0.01$, 0.980 58

8 $1 - \dfrac{3}{2}x^2 + \dfrac{27}{8}x^4 - \dfrac{135}{16}x^6$

9 $1 + \dfrac{x}{2} - \dfrac{x^2}{8} + \dfrac{x^3}{16}$, $\dfrac{1145}{512}$

10 a $n = -2,\ a = 3$
b -108
c $|x| < \frac{1}{3}$

11 a $\dfrac{3}{(1+x)} + \dfrac{6}{(2+x)} - \dfrac{4}{(2+x)^2}$
b $B = 3,\ C = \dfrac{-23}{8}$
c $|x| < 1$

Exercise 4A

1 a $\dfrac{2t-3}{2}$ **b** $\dfrac{6t^2}{6t} = t$ **c** $\dfrac{4}{1+6t}$
d $\dfrac{15t^3}{2}$ **e** $-3t^3$ **f** $t(1-t)$
g $\dfrac{2t}{t^2-1}$ **h** $\dfrac{2}{(t^2+2t)e^t}$ **i** $-\frac{3}{4}\tan 3t$
j $4\tan t$ **k** $\operatorname{cosec} t$ **l** $\cot t$

2 a $y = \dfrac{\pi}{6}x + \dfrac{\pi}{3}$ **b** $2y + 5x = 57$
3 a $x = 1$ **b** $y + \sqrt{3}x = \sqrt{3}$
4 $(0, 0)$ and $(-2, -4)$

Exercise 4B

1 a $-\dfrac{2x}{3y^2}$ **b** $-\dfrac{x}{5y}$ **c** $\dfrac{x+3}{4-5y}$
d $\dfrac{4-6xy}{3x^2+3y^2}$ **e** $\dfrac{3x^2-2y}{6y+2x-2}$ **f** $\dfrac{3x^2-y}{2+x}$
g $\dfrac{4(x-y)^3-1}{1+4(x-y)^3}$ **h** $\dfrac{e^y-ye^x}{e^x-xe^y}$ **i** $-\dfrac{(2\sqrt{xy}+y)}{(4y\sqrt{xy}+x)}$

2 $9y + 7x = 23$
3 $y = 2x - 2$
4 $(3, 1)$ and $(3, 3)$

Exercise 4C

1 a $3^x \ln 3$ **b** $\left(\frac{1}{2}\right)^x \ln\left(\frac{1}{2}\right)$
c $a^x(1 + x\ln a)$ **d** $\dfrac{2^x(x\ln 2 - 1)}{x^2}$

2 $4y = 15\ln 2(x-2) + 17$
3 -9.07 millicuries/day
4 $P = 37\,000\,k^t$ where $k = \sqrt[100]{\dfrac{109}{37}}$
1178 people per year
Rate of increase of population during the year 2000.

Exercise 4D

1 $\dfrac{8}{9\pi}$ **2** 6π **3** $15e^2$ **4** $-\frac{9}{2}$

Exercise 4E

1 $\dfrac{dM}{dt} = -kM$
4 $\dfrac{dQ}{dt} = -kQ$
5 $\dfrac{dx}{dt} = \dfrac{k}{x^2}$
6 $\dfrac{dP}{dt} = kP - Q$
7 $\dfrac{dr}{dt} = \dfrac{k}{r}$
8 $\dfrac{d\theta}{dt} = -k(\theta - \theta_0)$
11 $\dfrac{dh}{dt} = -\dfrac{18}{\pi h^2}$

Mixed exercise 4F

1 a $\dfrac{dy}{dx} = -\dfrac{4}{t^3}$ **b** $y = 2x - 8$
2 $3y + x = 33$
3 $y = \frac{2}{3}x + \frac{1}{3}$
4 a $\dfrac{dx}{dt} = -2\sin t + 2\cos 2t$; $\dfrac{dy}{dt} = -\sin t - 4\cos 2t$
b $\frac{1}{2}$
c $y + 2x = \dfrac{5\sqrt{2}}{2}$
5 b 2
6 $\dfrac{dV}{dt} = -kV$
7 a $-\frac{4}{5}$ **b** $5y + 4x = 20\sqrt{2}$ **c** $(5\sqrt{2}, 0)$
8 a $\dfrac{\pi}{6}$ **b** $-\frac{3}{16}\operatorname{cosec} t$ **d** $-\frac{123}{64}$
9 a $-\frac{1}{2}\sec t$ **b** $4y + 4x = 5a$
10 $y + x = 16$
11 a $\dfrac{\pi^2}{4}$ **b** $\dfrac{\cos 2t}{t}$
12 $\frac{1}{7}$
13 $\dfrac{y - 2e^{2x}}{2e^{2y} - x}$
14 $(1, 1)$ and $(-\sqrt[3]{-3}, \sqrt[3]{-3})$.
15 $-\frac{1}{4}$
16 a $\dfrac{2x - 2 - y}{1 + x - 2y}$
b $\frac{4}{3}, -\frac{1}{3}$
c $\left(\dfrac{5+2\sqrt{13}}{3}, \dfrac{4+\sqrt{13}}{3}\right)$ and $\left(\dfrac{5-2\sqrt{13}}{3}, \dfrac{4-\sqrt{13}}{3}\right)$
19 b $\dfrac{6t}{2^t \ln 2}$ **c** 1.307
21 a $\dfrac{\ln P - \ln P_0}{\ln 1.09}$ **b** 8.04 years **c** $0.172\,P_0$

Exercise 5A

1

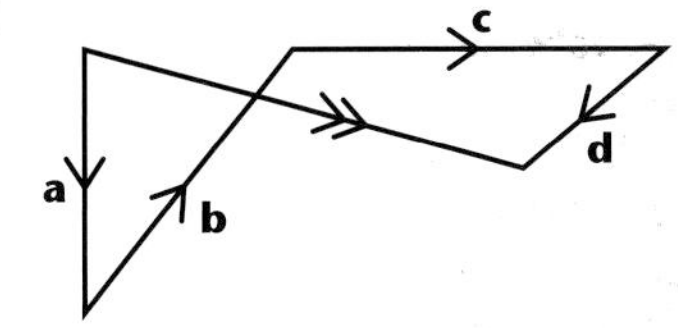

2 25
3 $\sqrt{569} \approx 23.9$
4 a **d** − **a** **b** **a** + **b** + **c**
c **a** + **b** − **d** **d** **a** + **b** + **c** − **d**

Exercise 5B

1 a 2**a** + 2**b** **b** **a** + **b** **c** **b** − **a**
2 a **b** − $\frac{1}{2}$**a** **b** **b** − 3**a**
c $\frac{3}{2}$**a** − **b** **d** 2**a** − **b**
3 a Yes ($\lambda = 2$) **b** Yes ($\lambda = 4$)
c No **d** Yes ($\lambda = -1$)
e Yes ($\lambda = -3$) **f** No
4 a $\lambda = \frac{1}{2},\ \mu = -3$ **b** $\lambda = -2,\ \mu = 1$
c $\lambda = \frac{1}{4},\ \mu = 5$ **d** $\lambda = -2,\ \mu = -1$
e $\lambda = 4,\ \mu = 8\frac{1}{2}$
5 a **b** − **a**, $\frac{5}{6}$(**b** − **a**), $\frac{1}{6}$**a** + $\frac{5}{6}$**b** **b** $-\frac{1}{6}$**a** + $(\lambda - \frac{5}{6})$**b**
c $-\mu$**a** + $(\mu - \lambda)$**b** **d** $\lambda = \frac{1}{2},\ \mu = \frac{1}{6}$
6 a i −**a** + **b** **ii** $\frac{2}{3}$**a** − $\frac{7}{4}$**b**
b $(\frac{2}{3}\lambda - \mu)$**a** + $(\frac{3}{4} - \frac{7}{4}\lambda + \mu)$**b** = **0**
d $\frac{6}{13}$
e $\frac{6}{13}$**a** + $\frac{7}{13}$**b**
f $\frac{13}{10}$

7 a $-\mathbf{a}+\mathbf{b}$ **b** $\frac{1}{2}\mathbf{a}+\frac{1}{2}\mathbf{b}$

c $\frac{3}{8}\mathbf{a}+\frac{3}{8}\mathbf{b}$ **d** $-\frac{5}{8}\mathbf{a}+\frac{3}{8}\mathbf{b}$

e $-\mathbf{a}+k\mathbf{b}$ **f** $5:3$, $k=\frac{3}{5}$

8 a $\frac{1}{3}\mathbf{a}$

b $\frac{1}{4}\mathbf{a}+\frac{3}{4}\mathbf{b}$

c $-\frac{1}{12}\mathbf{a}+\frac{3}{4}\mathbf{b}$

d $2\mathbf{b}$

e $-\frac{1}{4}\mathbf{a}+\frac{1}{4}\mathbf{b}$

f $-\frac{1}{4}\mathbf{a}+\frac{9}{4}\mathbf{b}$

g $1:3$

h $-\frac{1}{3}\mathbf{a}+\mathbf{b}$, $-\mathbf{a}+3\mathbf{b}$, $\overrightarrow{AG}=3\overrightarrow{EB}$ $\therefore$ parallel

Exercise 5C

1 $\frac{5}{6}\mathbf{a}+\frac{1}{6}\mathbf{b}$

2 $-\frac{1}{2}\mathbf{a}-\frac{1}{2}\mathbf{b}+\mathbf{c}$

3 $\overrightarrow{OC}=-2\mathbf{a}+2\mathbf{b}$, $\overrightarrow{OD}=-3\mathbf{a}+2\mathbf{b}$, $\overrightarrow{OE}=-2\mathbf{a}+\mathbf{b}$

Exercise 5D

1 a $\begin{pmatrix}12\\3\end{pmatrix}$ **b** $\begin{pmatrix}-1\\16\end{pmatrix}$ **c** $\begin{pmatrix}-21\\-29\end{pmatrix}$

2 a $3\mathbf{i}-\mathbf{j}$, $4\mathbf{i}+5\mathbf{j}$, $-2\mathbf{i}+6\mathbf{j}$

b $\mathbf{i}+6\mathbf{j}$

c $-5\mathbf{i}+7\mathbf{j}$

d $\sqrt{40}=2\sqrt{10}$

e $\sqrt{37}$

f $\sqrt{74}$

3 a $\frac{1}{5}\begin{pmatrix}4\\3\end{pmatrix}$ **b** $\frac{1}{13}\begin{pmatrix}5\\-12\end{pmatrix}$

c $\frac{1}{25}\begin{pmatrix}-7\\24\end{pmatrix}$ **d** $\frac{1}{\sqrt{10}}\begin{pmatrix}1\\-3\end{pmatrix}$

4 -7 or -23

Exercise 5E

1 $\sqrt{84}\approx 9.17$

2 $\sqrt{147}\approx 12.1$

3 a $\sqrt{14}\approx 3.74$ **b** 15

c $\sqrt{50}\approx 7.07$ **d** $\sqrt{30}\approx 5.48$

4 5 or 9

5 -4 or 10

Exercise 5F

1 a $\sqrt{35}$ **b** $\sqrt{20}=2\sqrt{5}$ **c** $\sqrt{3}$

d $\sqrt{170}$ **e** $\sqrt{75}=5\sqrt{3}$

2 a $\begin{pmatrix}7\\1\\-1\end{pmatrix}$ **b** $\begin{pmatrix}-5\\5\\-5\end{pmatrix}$ **c** $\begin{pmatrix}14\\-3\\1\end{pmatrix}$

d $\begin{pmatrix}8\\4\\4\end{pmatrix}$ **e** $\begin{pmatrix}8\\-6\\10\end{pmatrix}$ **f** $10\sqrt{2}$

3 $7\mathbf{i}-3\mathbf{j}+2\mathbf{k}$

4 6 or -6

5 $\sqrt{3}$ or $-\sqrt{3}$

6 a $\begin{pmatrix}2t-2\\-4\\2t\end{pmatrix}$ **b** $\sqrt{8t^2-8t+20}$

c $t=\frac{1}{2}$ **d** $3\sqrt{2}$

7 a $\begin{pmatrix}-t\\4-t\\-1\end{pmatrix}$ **b** $\sqrt{2t^2-8t+17}$

c $t=2$ **d** 3

Exercise 5G

1 $\frac{9}{2}$

2 a 2 **b** 17 **c** -6

d 20 **e** 0

3 a 55.5° **b** 94.8° **c** 87.4° **d** 79.0°

e 100.9° **f** 53.7° **g** 132.2° **h** 70.5°

4 a -10 **b** 5 **c** $2\frac{3}{5}$

d $-2\frac{1}{2}$ **e** -5 or 2

5 a 32.9° **b** 117.8°

6 a 20.5° **b** 109.9°

7 $\dfrac{2\sqrt{2}}{3}$

9 a $2\mathbf{a}.\mathbf{b}+\mathbf{a}.\mathbf{c}$ **b** $13+2\mathbf{a}.\mathbf{b}$ **c** $2|\mathbf{a}|^2-|\mathbf{b}|^2$

10 a $\mathbf{i}+2\mathbf{j}+\mathbf{k}$ **b** $3\mathbf{i}+2\mathbf{j}+3\mathbf{k}$ **c** $3\mathbf{i}+2\mathbf{j}+4\mathbf{k}$

11 64.7°, 64.7°, 50.6°

12 a $\sqrt{33}$, $\sqrt{173}$ **b** 29.1°

13 $\frac{2}{27}$

Exercise 5H

1 a $\mathbf{r}=\begin{pmatrix}6\\5\\-1\end{pmatrix}+t\begin{pmatrix}2\\-3\\-1\end{pmatrix}$ **b** $\mathbf{r}=\begin{pmatrix}2\\5\\0\end{pmatrix}+t\begin{pmatrix}1\\1\\1\end{pmatrix}$

c $\mathbf{r}=\begin{pmatrix}-7\\6\\2\end{pmatrix}+t\begin{pmatrix}3\\1\\2\end{pmatrix}$ **d** $\mathbf{r}=\begin{pmatrix}2\\0\\4\end{pmatrix}+t\begin{pmatrix}-3\\2\\1\end{pmatrix}$

e $\mathbf{r}=\begin{pmatrix}6\\-11\\2\end{pmatrix}+t\begin{pmatrix}0\\5\\-2\end{pmatrix}$

2 a 34.4 **b** 28 **c** 20.4

3 $\mathbf{r}=\begin{pmatrix}4\\-3\\8\end{pmatrix}+t\begin{pmatrix}0\\0\\1\end{pmatrix}$

4 a $\mathbf{r}=\begin{pmatrix}2\\1\\9\end{pmatrix}+t\begin{pmatrix}2\\-2\\-1\end{pmatrix}$ **b** $\mathbf{r}=\begin{pmatrix}-3\\5\\0\end{pmatrix}+t\begin{pmatrix}10\\-3\\2\end{pmatrix}$

c $\mathbf{r}=\begin{pmatrix}1\\11\\-4\end{pmatrix}+t\begin{pmatrix}4\\-2\\6\end{pmatrix}$ **d** $\mathbf{r}=\begin{pmatrix}-2\\-3\\-7\end{pmatrix}+t\begin{pmatrix}14\\7\\4\end{pmatrix}$

5 a $p=1$, $q=10$

b $p=-6\frac{1}{2}$, $q=-21$

c $p=-19$, $q=-15$

Exercise 5I

1 Yes, (8, 7, 2)

2 No

3 Yes, (16, 2, -14)

4 Yes, (3, 1, 7)

5 No

Exercise 5J

1 79.5°

2 40.7°

3 81.6°

4 72.7°

5 76.9°

6 b $\frac{29}{30}$

c (10, 7, 2) or (-8, 1, 2)

Mixed exercise 5K

1 a $3\mathbf{i}+4\mathbf{j}+5\mathbf{k}$, $\mathbf{i}+\mathbf{j}+4\mathbf{k}$

2 $-7\mathbf{i}-2\mathbf{j}-5\mathbf{k}$

3 AD: $\mathbf{r} = (6\mathbf{i} + 8\mathbf{j}) + t(4\mathbf{i} - 3\mathbf{j})$
BC: $\mathbf{r} = (9\mathbf{i} + 12\mathbf{j}) + s(\mathbf{i} + 3\mathbf{j})$
$\frac{22}{3}\mathbf{i} + 7\mathbf{j}$
4 $(1, 1, 2)$, $70.5°$
5 $4\mathbf{i} + 7\mathbf{j} - 5\mathbf{k}$
6 $-\mathbf{i} + 3\mathbf{j} + \mathbf{k}$
7 $\mathbf{r} = \begin{pmatrix} 1 \\ -1 \\ 3 \end{pmatrix} + t\begin{pmatrix} 0 \\ 3 \\ -1 \end{pmatrix}$, $\mathbf{i} + \mathbf{j} + \frac{7}{3}\mathbf{k}$
8 a $\frac{1}{5}$ or 1
b $-3\mathbf{i} + 3\mathbf{j} + 8\mathbf{k}$
c $82°$
9 a 5
b $3\mathbf{i} + 4\mathbf{j} + 2\mathbf{k}$
c $\mathbf{r} = \begin{pmatrix} 5 \\ 1 \\ 5 \end{pmatrix} + t\begin{pmatrix} 2 \\ -3 \\ 3 \end{pmatrix}$
10 b $\sqrt{11}$ **c** $35°$ **d** 1.9
11 a $\mathbf{r} = \begin{pmatrix} 5 \\ -1 \\ -1 \end{pmatrix} + t\begin{pmatrix} -1 \\ -1 \\ 2 \end{pmatrix}$ **d** $3\mathbf{i} - 3\mathbf{j} + 3\mathbf{k}$
12 a $\mathbf{r} = \begin{pmatrix} 9 \\ -2 \\ 1 \end{pmatrix} + t\begin{pmatrix} -3 \\ 4 \\ 5 \end{pmatrix}$ **b** $p = 6$, $q = 11$
c $39.8°$ **d** $\frac{36}{5}\mathbf{i} + \frac{2}{5}\mathbf{j} + 4\mathbf{k}$
13 a $\mathbf{r} = \begin{pmatrix} 1 \\ 2 \\ -3 \end{pmatrix} + t\begin{pmatrix} 4 \\ -5 \\ 3 \end{pmatrix}$
c $19.5°$
d 1
14 b $5\mathbf{i} - \mathbf{k}$ **d** 1.5 km

Exercise 6A

1 a $3\tan x + 5\ln|x| - \frac{2}{x} + C$
b $5e^x + 4\cos x + \frac{x^4}{2} + C$
c $-2\cos x - 2\sin x + x^2 + C$
d $3\sec x - 2\ln|x| + C$
e $5e^x + 4\sin x + \frac{2}{x} + C$
f $\frac{1}{2}\ln|x| - 2\cot x + C$
g $\ln|x| - \frac{1}{x} - \frac{1}{2x^2} + C$
h $e^x - \cos x + \sin x + C$
i $-2\operatorname{cosec} x - \tan x + C$
j $e^x + \ln|x| + \cot x + C$
2 a $\tan x - \frac{1}{x} + C$
b $\sec x + 2e^x + C$
c $-\cot x - \operatorname{cosec} x - \frac{1}{x} + \ln|x| + C$
d $-\cot x + \ln|x| + C$
e $-\cos x + \sec x + C$
f $\sin x - \operatorname{cosec} x + C$ **g** $-\cot x + \tan x + C$
h $\tan x + \cot x + C$ **i** $\tan x + e^x + C$
j $\tan x + \sec x + \sin x + C$

Exercise 6B

1 a $-\frac{1}{2}\cos(2x + 1) + C$ **b** $\frac{3}{2}e^{2x} + C$
c $4e^{x+5} + C$ **d** $-\frac{1}{2}\sin(1 - 2x) + C$
e $-\frac{1}{3}\cot 3x + C$ **f** $\frac{1}{4}\sec 4x + C$
g $-6\cos(\frac{1}{2}x + 1) + C$ **h** $-\tan(2 - x) + C$
i $-\frac{1}{2}\operatorname{cosec} 2x + C$ **j** $\frac{1}{3}(\sin 3x + \cos 3x) + C$
2 a $\frac{1}{2}e^{2x} + \frac{1}{4}\cos(2x - 1) + C$
b $\frac{1}{2}e^{2x} + 2e^x + x + C$
c $\frac{1}{2}\tan 2x + \frac{1}{2}\sec 2x + C$
d $-6\cot(\frac{1}{2}x) + 4\operatorname{cosec}(\frac{1}{2}x) + C$
e $-e^{3-x} + \cos(3 - x) - \sin(3 - x) + C$
3 a $\frac{1}{2}\ln|2x + 1| + C$ **b** $-\frac{1}{2(2x + 1)} + C$
c $\frac{(2x + 1)^3}{6} + C$ **d** $\frac{3}{4}\ln|4x - 1| + C$
e $-\frac{3}{4}\ln|1 - 4x| + C$ **f** $\frac{3}{4(1 - 4x)} + C$
g $\frac{(3x + 2)^6}{18} + C$ **h** $\frac{3}{4(1 - 2x)^2} + C$
i $\frac{1}{(3 - 2x)^3} + C$ **j** $-\frac{5}{2}\ln|3 - 2x| + C$
4 a $-\frac{3}{2}\cos(2x + 1) + 2\ln|2x + 1| + C$
b $\frac{1}{5}e^{5x} - \frac{(1 - x)^6}{6} + C$
c $-\frac{1}{2}\cot 2x + \frac{1}{2}\ln|1 + 2x| - \frac{1}{2(1 + 2x)} + C$
d $\frac{(3x + 2)^3}{9} - \frac{1}{3(3x + 2)} + C$

Exercise 6C

1 a $-\cot x - x + C$
b $\frac{1}{2}x + \frac{1}{4}\sin 2x + C$
c $-\frac{1}{8}\cos 4x + C$
d $\frac{3}{2}x - 2\cos x - \frac{1}{4}\sin 2x + C$
e $\frac{1}{3}\tan 3x - x + C$
f $-2\cot x - x + 2\operatorname{cosec} x + C$
g $x - \frac{1}{2}\cos 2x + C$
h $\frac{1}{8}x - \frac{1}{32}\sin 4x + C$
i $-2\cot 2x + C$
j $\frac{3}{2}x + \frac{1}{8}\sin 4x - \sin 2x + C$
2 a $\tan x - \sec x + C$
b $-\cot x - \operatorname{cosec} x + C$
c $2x - \tan x + C$
d $-\cot x - x + C$
e $-2\cot x - x - 2\operatorname{cosec} x + C$
f $2\tan x - x + 2\sec x + C$
g $-\cot x - 4x + \tan x + C$
h $x + \frac{1}{2}\cos 2x + C$
i $-\frac{3}{2}x + \frac{1}{4}\sin 2x + \tan x + C$
j $-\frac{1}{2}\operatorname{cosec} 2x + C$
3 a $\frac{1}{6}\sin 3x + \frac{1}{2}\sin x + C$
b $-\frac{1}{8}\cos 8x - \frac{1}{2}\cos 2x + C$
c $-\frac{1}{8}\cos 8x + \frac{1}{2}\cos 2x + C$
d $-\frac{1}{7}\sin 7x + \frac{1}{3}\sin 3x + C$
e $\frac{1}{5}\sin 10x + \frac{1}{2}\sin 4x + C$
f $x + \frac{1}{8}\sin 8x + C$
g $-\frac{1}{8}\cos 8x + C$
h $x - \frac{1}{8}\sin 8x + C$

Exercise 6D

1 a $\ln|(x + 1)^2(x + 2)| + C$
b $\ln|(x - 2)\sqrt{2x + 1}| + C$
c $\ln\left|\frac{(x + 3)^3}{x - 1}\right| + C$

d $\ln\left|\dfrac{2+x}{1-x}\right| + C$

e $\ln\left|\dfrac{2x+1}{1-2x}\right| + C$

f $\frac{1}{3}\ln\left|\dfrac{(3x-1)^2}{3x+1}\right| + C$

g $\ln\left|\dfrac{(2-3x)^{\frac{1}{3}}}{(1-x)^2}\right| + C$

h $\ln|2+x| + \dfrac{2}{x+1} + C$

i $\ln\left|\dfrac{x+1}{x+2}\right| - \dfrac{2}{x+1} + C$

j $\ln\left|\dfrac{3+2x}{2-x}\right| + \dfrac{1}{2-x} + C$

2 a $x + \ln|(x+1)^2\sqrt{2x-1}| + C$

b $\dfrac{x^2}{2} + x + \ln\left|\dfrac{x^2}{(x+1)^3}\right| + C$

c $x + \ln\left|\dfrac{x-2}{x+2}\right| + C$

d $-x + \ln\left|\dfrac{(3+x)^2}{1-x}\right| + C$

e $-\dfrac{3}{x} - \ln|x+2| + C$

Exercise 6E

1 a $\frac{1}{2}\ln|x^2+4| + C$ **b** $\frac{1}{2}\ln|e^{2x}+1| + C$

c $-\frac{1}{4}(x^2+4)^{-2} + C$ **d** $-\frac{1}{4}(e^{2x}+1)^{-2} + C$

e $\frac{1}{2}\ln|3+\sin 2x| + C$ **f** $\frac{1}{4}(3+\cos 2x)^{-2} + C$

g $\frac{1}{2}e^{x^2} + C$ **h** $\frac{1}{10}(1+\sin 2x)^5 + C$

i $\frac{1}{3}\tan^3 x + C$ **j** $\tan x + \frac{1}{3}\tan^3 x + C$

2 a $\frac{1}{10}(x^2+2x+3)^5 + C$ **b** $-\frac{1}{4}\cot^2 2x + C$

c $\frac{1}{18}\sin^6 3x + C$ **d** $e^{\sin x} + C$

e $\frac{1}{2}\ln|e^{2x}+3| + C$ **f** $\frac{1}{5}(x^2+1)^{\frac{5}{2}} + C$

g $\frac{2}{3}(x^2+x+5)^{\frac{3}{2}} + C$ **h** $2(x^2+x+5)^{\frac{1}{2}} + C$

i $-\frac{1}{3}(\cos 2x+3)^{\frac{1}{2}} + C$ **j** $-\frac{1}{4}\ln|\cos 2x+3| + C$

Exercise 6F

1 a $\frac{2}{5}(1+x)^{\frac{5}{2}} - \frac{2}{3}(1+x)^{\frac{3}{2}} + C$

b $\frac{2}{3}(1+x)^{\frac{3}{2}} - 2\sqrt{1+x} + C$

c $-\ln|1-\sin x| + C$

d $\dfrac{(3+2x)^7}{28} - \dfrac{(3+2x)^6}{8} + C$

e $\dfrac{\cos^3 x}{3} - \cos x + C$

2 a $\frac{2}{5}(2+x)^{\frac{5}{2}} - \frac{4}{3}(2+x)^{\frac{3}{2}} + C$

b $\ln\left|\dfrac{\sqrt{x}-2}{\sqrt{x}+2}\right| + C$

c $\frac{2}{5}(1+\tan x)^{\frac{5}{2}} - \frac{2}{3}(1+\tan x)^{\frac{3}{2}} + C$

d $\sqrt{x^2+4} + \ln\left|\dfrac{\sqrt{x^2+4}-2}{\sqrt{x^2+4}+2}\right| + C$

e $\tan x + \frac{1}{3}\tan^3 x + C$

3 a $\frac{506}{15}$ or 33.73 **b** $\frac{16}{3} - 2\sqrt{3}$ or 1.87

c $2 + 2\ln\frac{2}{3}$ or 1.19 **d** $2 - 2\ln 2$ or 0.614

e 9.7 **f** $\frac{1}{2}\ln\frac{9}{5}$ or 0.294

Exercise 6G

1 a $-x\cos x + \sin x + C$

b $x e^x - e^x + C$

c $x\tan x - \ln|\sec x| + C$

d $x\sec x - \ln|\sec x + \tan x| + C$

e $-x\cot x + \ln|\sin x| + C$

2 a $\dfrac{x^3}{3}\ln x - \dfrac{x^3}{9} + C$

b $3x\ln x - 3x + C$

c $-\dfrac{\ln x}{2x^2} - \dfrac{1}{4x^2} + C$

d $x(\ln x)^2 - 2x\ln x + 2x + C$

e $\dfrac{x^3}{3}\ln x - \dfrac{x^3}{9} + x\ln x - x + C$

3 a $-e^{-x}x^2 - 2x e^{-x} - 2e^{-x} + C$

b $x^2\sin x + 2x\cos x - 2\sin x + C$

c $x^2(3+2x)^6 - \dfrac{x(3+2x)^7}{7} + \dfrac{(3+2x)^8}{112} + C$

d $-x^2\cos 2x + x\sin 2x + \frac{1}{2}\cos 2x + C$

e $x^2\sec^2 x - 2x\tan x + 2\ln|\sec x| + C$

4 a $2\ln 2 - \frac{3}{4}$ **b** 1

c $\dfrac{\pi}{2} - 1$ **d** $\frac{1}{2}(1-\ln 2)$

e 9.8 **f** $2\sqrt{2}\pi + 8\sqrt{2} - 16$

g $\frac{1}{2}(1-\ln 2)$

Exercise 6H

1 a 3.42 **b** 1.34 **c** 1.04

d 2.42 **e** 1.41

2 a $8\ln 4 - \frac{15}{4}$

b i 7.45 **ii** 7.37

c i 1.6% **ii** 0.4%

3 a i 1.509 **ii** 1.329 **iii** 1.282

b Halving h reduces differences by $(\frac{1}{3})$

$0.18 \rightarrow 0.05 \rightarrow$ possibly 0.02 suggests 1.25–1.27

4 a $\frac{16}{15}\sqrt{2}$

b i 1.34 **ii** 1.42

c i 11.4% **ii** 6.1%

Increasing the number of strips improves accuracy.

Exercise 6I

1 a i $2\ln 2$ **ii** 2π

b i $\ln(2+\sqrt{3})$ **ii** $\sqrt{3}\pi$

c i $2\ln 2 - 1$ **ii** $[2(\ln 2)^2 - 4\ln 2 + 2]\pi$

d i $\sqrt{2} - 1$ **ii** $\frac{1}{3}\pi$

e i $\frac{8}{3}$ **ii** $\frac{64}{15}\pi$

2 a $\ln 4$ **b** $\ln 3 - \frac{2}{3}$ **c** 1

d $\dfrac{(2\sqrt{2}-1)}{3}$ **e** $\frac{1}{2}(1-\ln 2)$

3 a $\frac{192}{5}$ **b** $\frac{768}{7}\pi$

4 a $\frac{2}{3}$ **b** $\frac{8}{15}\pi$

Exercise 6J

1 a $y = A e^{x - x^2} - 1$ **b** $y = k\sec x$

c $y = \dfrac{-1}{\tan x - x + C}$ **d** $y = \ln(2e^x + C)$

e $y = Ax e^{-\frac{1}{x}}$

2 a $\sin y = k\sec x$

b $\ln|\sec y| = x\tan x - \ln|\sec x| + C$

c $\dfrac{1+y}{1-y} = k(1+x^2)$

d $\cos 2y = 2\cot x + C$

e $\ln|2+e^y| = -x e^{-x} - e^{-x} + C$

3 a $\ln|y| = e^x + C$

b $-e^{-y} = \frac{x^2}{2} + C$

c $\ln y = \sin x + C$ or $y = A\,e^{\sin x}$

d $\ln|\sec y + \tan y| = \frac{x^2}{2} + C$

e $\ln|\sec y + \tan y| = x + \frac{1}{2}\sin 2x + C$

f $\frac{1}{2}\tan y = \sin x + C$

4 a $\frac{1}{24} - \frac{\cos^3 x}{3}$

b $\sin 2y + 2y = 4\tan x - 4$

c $\tan y = \frac{1}{2}\sin 2x + x + 1$

d $y = \frac{3}{1-x}$

e $\frac{1+y}{1-y} = \frac{1+x}{2}$

Exercise 6K

1 $3e^4$

2 $3\ln 2$

3 a $M = \frac{e^t}{1+e^t}$

b $\frac{2}{3}$

c M approaches 1

4 235

5 $46\frac{2}{3}$

6 $\frac{2}{k}$

7 4

Mixed exercise 6L

1 a $x = 4, y = 20$ **c** $\frac{62}{5} + 48\ln 4$

2 a $y = 2x - 8$ **c** 16.2

3 b $\pi(\frac{3}{4}\pi + 2)$

4 a $\frac{1}{2}e^{\frac{1}{2}x} - \frac{1}{x^2}$

b $f'(1.05) = -0.06$, $f'(1.10) = +0.04$; change of sign, $\therefore$ root

c $2e^{\frac{1}{2}x} + \ln|x| + c$

d 10.03

5 a $-x\,e^{-x} - e^{-x}$ **b** $\cos 2y = 2(x\,e^{-x} + e^{-x} - 1)$

6 $\frac{56\pi}{5}$

7 a $\ln|x| - \ln|x+1| + k_1$

b $\ln e^x - \ln(e^x + 1) + k_2$, where k_1 and k_2 are arbitrary constants.

c $-x^2\cos x + 2x\sin x + 2\cos x + C$

8 a $-\frac{1}{2}x\cos 2x + \frac{1}{4}\sin 2x$

b $\tan y = -\frac{1}{2}x\cos 2x + \frac{1}{4}\sin 2x - \frac{1}{4}$

9 i $\frac{1}{2}x\sin 2x + \frac{1}{4}\cos 2x + c$ **ii** $\pi\left(\frac{\pi^2}{4} + 1\right)$

10 a $f(x) = \frac{1}{4}e^{2x} - 3x^2 - \frac{5}{4}$

11 a $\frac{1}{4}$

b $\frac{32}{3}x^{\frac{3}{2}} - 2\ln|x| + C$

c $\frac{224}{3} - 2\ln 4$

12 118.4π

13 a $\frac{1}{12}(x^2+3)^6 + C$

14 a $A = \frac{1}{2}, B = 2, C = -1$

b $\frac{1}{2}\ln|x| + 2\ln|x-1| + \frac{1}{x-1} + C$

15 a $-\frac{4}{5}$ **b** $y - 2\sqrt{2} = -\frac{4}{5}\left(x - \frac{5}{\sqrt{2}}\right)$

c $(5\sqrt{2}, 0)$ **d** $10 - 2.5\pi$

16 a $-\frac{1}{y} = \frac{x^2}{2} + C$ **c** 1

e $(-2, -2)$

17 a $\frac{dy}{dx} = \frac{-a\sin t}{2a\sin t\cos t} = \frac{-1}{2\cos t}$

b $y + x = \frac{5}{4}a$

d $\frac{1}{96}\pi a^3$

18 a $\ln|1+2x| + \frac{1}{1+2x} + C$

b $2y - \sin 2y = \ln|1+2x| + \frac{1}{1+2x} + \frac{\pi}{2} - 2$

19 $A_1 = \frac{1}{4} - \frac{1}{2e}$, $A_2 = \frac{1}{4}$

20 $e^{-x}(-x^2 - 2x - 2) + C$

$\frac{1}{3}e^{-3y} = e^{-x}(x^2 + 2x + 2) - \frac{5}{3}$

21 a $\frac{1}{3}\ln 7$

22 a $A = 1, B = \frac{1}{2}, C = -\frac{1}{2}$

b $x + \frac{1}{2}\ln|x-1| - \frac{1}{2}\ln|x+1| = 2t - \frac{1}{2}\ln 3$

23 a $y = -\frac{1}{3}x + 11$

b

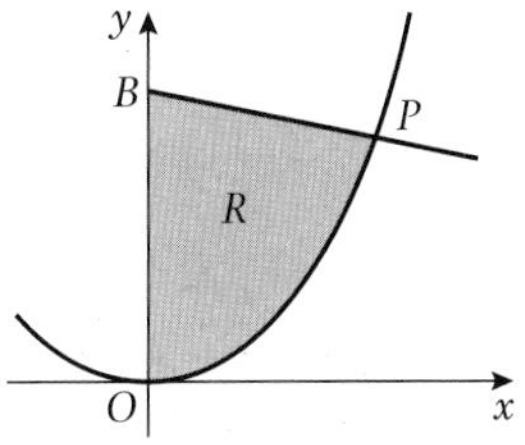

c 186π

24 $\frac{1}{2}e^2 + e^{-1} - 1\frac{1}{2}$

25 b $\frac{1}{2}(x - \sin x\cos x) + C_1$

c $\frac{1}{8}(2x^2 - 4x\sin x\cos x - \cos 2x) + C_2$, where C_1 and C_2 are arbitrary constants of integration.

26 b

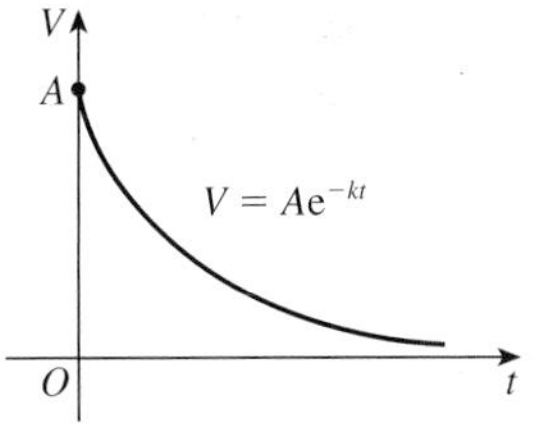

27 a $x + y = 16$ **b** 58.9

28 a $\frac{\pi^2}{4}$ **b** $\frac{\cos 2t}{t}$ **d** $\frac{\pi}{2}$

29 a $\frac{1}{3}\sin^3 x + C$ **b** $\frac{x^2}{2}\ln x - \frac{1}{4}x^2 + C$

c $\frac{2}{3}(x-2)\sqrt{x+1} + C$ **d** $\frac{8}{3}$

30 a $\frac{(1+2x^2)^6}{24} + C$ **b** $\tan 2y = \frac{1}{12}(1+2x^2)^6 + \frac{11}{12}$

31 $\frac{x^3}{3}\ln 2x - \frac{x^3}{9} + C$

32 $y^2 = \frac{8x}{x+2}$

33 b 1.38 **c** 2.05

34 a $y^2 = 4x^2(9 - x^2)$ **b** $A = 27$

c 18 **d** 36 cm^2

Examination style paper

1 $\frac{1}{4} - \frac{x}{4} + \frac{3x^2}{16} - \frac{x^3}{8}$

2 $\frac{10}{3}$

3 $\frac{5}{108}$

4 **a** $A = -\frac{2}{5}$, $B = \frac{1}{5}$

b $\ln k\left(\frac{x-2}{2x+1}\right)^{\frac{1}{5}}$

c $y = 49\left(\frac{x-2}{2x+1}\right)^2$

5 **a** $\frac{dP}{dt} = kP, k > 0$

c 10.8 million (3 sf)

6 **a** $\mathbf{r} = \mathbf{i} - 5\mathbf{j} - 7\mathbf{k} + t(9\mathbf{i} + 15\mathbf{j} + 12\mathbf{k})$

b $4\mathbf{i} - 3\mathbf{k}$

c $\frac{75\sqrt{2}}{2}$

d 1 : 2

7 **a** $\frac{2}{3}$

c $5 - 3\sqrt{3} + \frac{3\pi}{2}$

Index

addition
simple fractions 1–2, 1–2
vectors 48–49
angle between two straight lines 76–77
answers to exercises 125–134
area of region, integration 105–108
area under a curve, parametric equations 16–19

binomial expansion
$(1 + x)^n$ n is a positive integer 21–25
$(1 + x)^n$ n is not a positive integer 22–23
$(a + bx)^n$ 26–28
complex expressions 28–30
partial fractions 28–30
summary of key points 32
boundary conditions 109

cartesian equations
converting from parametric equations 14–15
of curve from parametric equation 10–11
chain rule 16, 33, 35, 38
chain rule reversed 84–86
column matrix 57, 62–63
connected rate of change 38–39, 42
cooling rate example 40
coordinate geometry, summary of key points 20
curve, cartesian equation from parametric 10–11

differential equations
information given in context 39–42
integration 108–112
differentiation
chain rule 33, 35, 38
gradient of a curve in parametric coordinates 33–34
implicit 97
implicit equations 35–36
power function a^x 37
product rule 35, 36, 99
rate of change of growth or of decay 37–39
summary of key points 46

equating coefficients to create partial fractions 2–3
examination style paper 123–124, answers 134

fractions
improper 6–7
mixed number 6–7
simple 1–2
splitting into partial fractions 2–3, 3–4, 5
with more than two linear factors 3–4
with repeated linear terms 5
with two linear factors 2–3

gradient of a curve, parametric coordinates 33–34

hemisphere area and volume 38

implicit equations, differentiation 35–36
improper fractions into partial fractions 6–7
information given in context 39–42
integration
area of region 105–108
boundary conditions 109
by parts 99–102
by substitution 95–99
chain rule reversed 84–86
changing the variable 95–99
differential equations 108–112
general pattern of expressions 93
implicit differentiation 97
linear transformations of some functions 84–86
modulus sign 82, 91
numerical 102–104
partial fractions 89–92
quotient rule 85
separation of variables 108–110
standard functions 82–83
standard functions families 92–94
standard functions generalised 85
standard trigonometric functions 102
summary of key points 121–122
trapezium rule 102–104
trigonometric functions 102
trigonometric identities 86–89
volume of revolution 105–108
intersection of two straight lines 74–76

lowest common multiple 1

mixed number fraction into partial fractions 6–7
modulus of vectors 57, 62–63

Newton's Law of Cooling 40
notation
letters for scalar parameters 74
modulus of vector 49–50
modulus sign in integration 82, 91
position vector 55
scalar product 64
vectors – printing 48
vectors – writing 48
numerical integration 102–104

parametric coordinates, gradient of a curve 33–34
parametric equations
area under a curve 16–19
conversion into cartesian equations 14–15
introduction 9–11
point on a curve 9–11
problem solving 11–13
partial fractions
binomial expansion 28–30
created by equating coefficients 2–3
created by substitution 2–3
fraction with more than two linear factors 3–4
fraction with repeated linear terms 5
fraction with two linear factors 2–3
integration 89–92
summary of key points 8
point on a curve parametric equation 9–11
point, position of using vectors 55–56
population growth example 40
position vector 55–56
product rule
differentiation 35, 36, 99
Pythagoras' Theorem 3–D version 60–61

quotient rule 85

rate of change of growth or of decay 37–39
rate of change, connected 38–39, 42

scalar parameter 71, 72
scalar product 64–70
scalar product in cartesian component form 66–67
scalar quantity 47
separation of variables 108–110
simple fractions
  addition 1–2
  subtraction 1–2
straight line equations 70–74
substitution to create partial fractions 2–3
subtraction
  simple fractions 1–2
  vectors 51
summaries of key points
  binomial expansion 32
  coordinate geometry 20
  differentiation 46
  integration 121–122
  partial fractions 8
  vectors 80–81
surds, use for exactness 58

trapezium rule 102–104
triangle law 48–49
trigonometric functions, integration 102
trigonometric identities, integration 86–89

unit vector 51–52, 62–63

vector quantity 47
vectors
  addition 48–49
  algebra of scalar products 65
  angle between two 64–70
  angle between two straight lines 76–77
  arithmetic 51–54
  cartesian components in 2–D 56–59
  cartesian coordinates in 3–D 59–61
  column matrix 57, 62–63
  equal 48
  intersection of two straight lines 74–76
  modulus 49–50, 57, 62–63
  non-zero, non-parallel 52
  notation 48, 49, 64, 74
  parallel 51
  perpendicular non-zero 65, 68
  position of a point 55–56
  scalar parameter 71, 72
  scalar product 64–70
  scalar product in cartesian component form 66–67
  scalar quantity 47
  straight line equations 70–74
  subtraction 51
  summary of key points 80–81
  3–D usage 62–63
  triangle law 48–49
  unit 51–52, 62–63
  vector quantity 47
  zero 49
volume loss by evaporation example 40–41
volume of revolution, integration 105–108

zero vector 49